Society and Its Environment

Society and Its Environment

An Introduction

Egbert Tellegen and Maarten Wolsink
University of Amsterdam, The Netherlands

Routledge
Taylor & Francis Group
LONDON AND NEW YORK

Published by Routledge
2 Park Square, Milton Park, Abingdon, Oxon, OX14 4RN
270 Madison Ave, New York NY 10016

Transferred to Digital Printing 2006

Originally published in Dutch in 1994 as Milieu en Samenleving: een Sociologische inleiding by Stenfert Kroese

---

British Library Cataloguing in Publication Data

ISBN : 90-5699-126-4 (Soft)

Publisher's Note
The publisher has gone to great lengths to ensure the quality of this reprint but points out that some imperfections in the original may be apparent

Contents

Acknowledgements

This book is based on the work of many other scientists. Their publications are listed in the references at the end of this book.
We have used the writings and figures of many others for preparing our own text and graphs. Fragments of texts (quoted in frame), figures or tables from several authors and organizations have been reproduced. Anyone using their work based on the quotes and references in this book is kindly requested to give the original authors full credit. Those whose work we have reprinted are mentioned below, together with the pages of this book where their work appears.
We are grateful to the authors and publishers who gave us permission to make use of their work.

R. Behnke and C. Kerven (76); W. Biesiot *et al.* (119); M. Bookchin (67); T. Brenton (233); D.B. Brobst (7); K.M. Cleaver and G.A. Schreiber (45); B. Commoner (19); R.R. Costanza *et al.* (15); M. Couto Soares *et al.* (235); H. Dieleman *et al.* (182); The Ecologist (68); G. Hardin (92); R. Heilbroner (27; 30); F. Hirsch (35); E.J. Hobsbawm (29); O. Hohmeyer (87); T.F. Homer-Dixon (54); M. Jänicke *et al.* (40); F. Langeweg *et al.* (50); P.M. Lehmann (46); Lucretius (23); R. Malthus (16); K. Marx (17; 65); C. Merchant (63); J.S. Mill (34); Netherlands Central Bureau of Statistics (13; 18); OECD (31); G. Prunier (47); L. Reijnders (180); J.J. Rousseau (63); A. Rowell (177); S. Schama (70); L. Schipper and S. Meyer (40); M. Schwartz and M. Thompson (78); V. Smil (44); D. Stanners and Ph. Bourdeau (230); D.P. Sutton and N.P. Harmon (10); H. Udo de Haes (5); UNEP (9); S. Visser (31); M. Weber (80); G.M. Woodwell (11); B. Wynne (164).

Introduction

People cannot live without changing nature. They do so by breathing, feeding and defecating, by dressing and heating and by creating barriers against wind and water, cold and heat. These forms of human-induced change of nature have been present since the dawn of mankind. People are constantly confronted by a malignant nature against which they have to defend themselves and whose resources they have to use in order to survive.

When people adapt nature to suit their own interests, they may come into conflict with other people having other interests regarding the state of the surrounding nature. With some exaggeration one can say that men cannot live without changing nature nor without quarrelling about these changes. This has been the human condition since pre-historic times.

However, the relationship between man and nature has dramatically changed during the past centuries. On the one hand the number of people on earth and the extent and diversity of the human use of natural resources has drastically changed. On the other hand, people have become more concerned about the damaging, if not catastrophic, effects of human-induced changes of nature than ever before. More than the word 'nature', the term 'environment' has become strongly associated with damage and decay caused by human beings. 'Environment' itself may be a neutral term, it is generally used within the context of environmental issues. Whereas the word 'nature' is often associated with 'beauty', 'silence', 'adventure' and other positive experiences, the term 'environment' is mostly used within the context of developments which are considered problematic. Hence, in practice 'environment' is mostly associated with problems. In this book the term 'environment' is used with regard to this association. We do not describe different 'environments' and the way they are changed by human activities in general, but we focus on those human-induced environmental changes which are considered as problematic by society or parts of it. What are the causes of these changes, when and where are these changes considered as environmental problems and how do people react to these changes are the main questions of this book. One of the possible reactions to environmental problems is the efforts to solve them. The ways in which individual citizens, private enterprises, public authorities, environmental organizations and others try to solve environmental problems is a main topic of this book.

The term 'environment' refers to both the natural and man-made physical surroundings of human life. As such, the study of environmental change is a task of natural scientists. However, both human causes of environmental change and human reactions to these changes are objects of study within the social and behavioural sciences. Why and how people change their environment, and when and where they react to these changes are highly relevant questions for social scientists. By the choice of the title 'Society and its environment', we want to express the two main social characteristics of environmental problems. They are consequences of human behaviour and at the same time a result of human perception.

In the first chapter we introduce and discuss the three elements of environmental problems: environmental change, caused by human behaviour and perceived as problematic by human beings.

The following two chapters discuss environmental problems in their temporal and spatial dimensions. Our discussion of culture and civilization in Chapter 4 is written much more from a temporal perspective than from a spatial perspective. Its basic question is whether and how today's environmental problems are related to the development of western culture and civilization.

After the presentation of a great variety of environmental problems in different times and at different places, Chapter 5 offers a more analytical approach to environmental problems, starting with the transfer of disadvantages as the main characteristic of environmental problems and discussing the ways in which people react to the dilemmas of individual benefits versus costs that are transferred to a collective.

In the second half of the book we present environmental problems at different levels of social scale. In Chapters 6 and 7 individuals are the starting point. In most chapters of the book both causes and perceptions of environmental problems are discussed. However, in Chapters 6 and 7 they are rather sharply separated. Chapter 6 on attitudes and behaviour mainly deals with human causation of environmental change and Chapter 7 discusses the human perceptions of incidental and constant human-induced environmental changes and reactions to these changes.

In Chapter 8 our subject is organizations. It would be beyond the scope of an introductory book like this to discuss all possible relationships between organizations and environmental problems. Instead of a general approach we have chosen a problem-solving approach. Organizations, in particular private enterprises, are often presented as main causes of environmental problems. However, in this chapter we discuss the ways in which different types of organizations such as environmental organizations, private enterprises and public utilities contribute to environmental problem-solving. Whereas 'organizations' are often considered as the bad guys, the state is often expected to be the good guy. Again our approach is different. Although it is stressed in Chapter 9 that the role of the state in solving environmental issues is crucial, a large part of this chapter on state and policy explains the contributions of states to the causation and continuation of environmental problems and the limitations of state environmental policies.

Our final chapter deals with the much discussed process of globalization. The main part of the chapter consists of a discussion of three contemporary global environmental issues: ozone depletion, climate change and reduction of biodiversity. Again, causation and construction of these problems are discussed.

This book is written by social scientists, more precisely by a sociologist and a political scientist. We have used studies from widely different social sciences such as history and geography, social psychology and political science, social philosophy and sociology, and also economics. However, we do not pretend to present these and other social sciences in a systematic way. This is a book about environmental issues and not about social environmental sciences. Our ambition is to introduce the reader to the social aspects of environmental problems in general and in different spatial and temporal contexts as well as at different levels of social scale.

In 1992 we published our book 'Milieu en samenleving' (Environment and society) in Dutch. This book is partly based on this earlier Dutch edition. However, only two chapters may be considered slightly adapted translations. In addition, the chapters on Culture and Globalization are completely new and the other chapters have been drastically changed.

We could not have written this book without the support of many people of whom we can mention only a few. We would like to thank Lucas Reijnders and Henk van Zon for their critical comments on earlier versions of several chapters. Ante Matser never failed to provide us with requested articles and books. Linda Pietersen improved the use of the English language in most of the chapters. In England Mrs Mary Boorman was of great help to us by editing and improving the Dutch–English translation. Elisabeth Segers and Marie-El Thunnissen helped us in preparing some of the graphs and tables.

Chapter 1

Environment and environmental problems

1.1 Interaction between 'environment' and 'society'

According to the dictionary, the concept of environment refers to all sorts of surroundings and includes material, social and spiritual conditions of living beings. However, in this book the term 'environment' only refers to non-human, living and non-living, physical surroundings. Characteristics of ecosystems, biodiversity, quality of air, water and soil, and stocks of non-renewable resources are included in our concept of the environment. Types of family relationships, conflicting or harmonious relationships in organisations and political systems are excluded from it.

In this book, therefore, the 'environment' is defined as a physical entity. As such it is an object of study by natural scientists. However, the (physical) environment continually changes under the influence of human behaviour. These changes are evaluated and reacted upon by human beings. The term 'society' in the title of this book refers to both human interventions which cause changes in the -physical- environment and human perceptions and interpretations of these changes (See figure 1.1.).

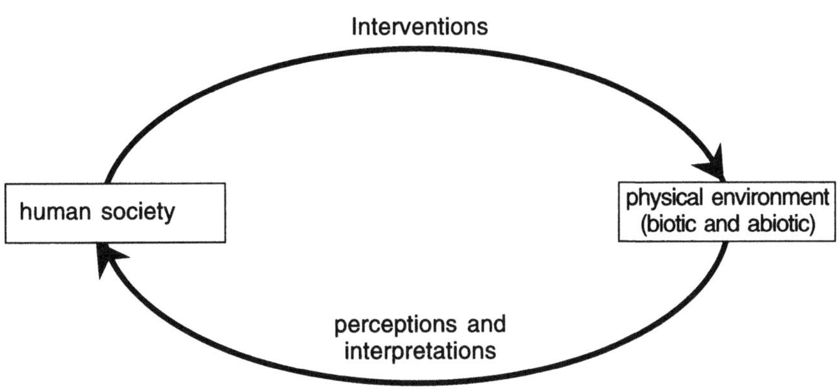

Figure 1.1 The interactions between man and environment (Udo de Haes, 1991, p.22).

1.2 Environmental problems

Unlike the concept of 'nature', which is strongly associated with positive qualities like beauty, purity and rarity, the concept of 'environment' is often used in the context of 'environmental problems', such as air pollution or destruction of tropical forests. Environmental problems are consequences of human intervention in the environment, as perceived and interpreted by human beings (see figure 1.1.). They may be defined as 'changes of the environment which are both caused by and considered as problematic by human beings'. The human causation of environmental problems may be clarified by the difference between natural and environmental catastrophes. We speak about 'natural catastrophes' when we assume that a dramatic change of the environment was caused by a natural process without human behaviour being involved. However, when speaking about environmental catastrophes we have dramatic events in mind which were, at least partly, caused by human activities. This is a analytical distinction. In concrete cases of floods, droughts and even earthquakes it may be uncertain whether or not human behaviour caused, or at least contributed to, the catastrophe. Nevertheless, for analytical purposes the distinction between natural and environmental catastrophes is extremely useful. We will not discuss interaction between man and environment in a general way, but focus on problematic forms of interactions between man and the environment.

In section 1.3 we will examine different types of human-induced environmental changes, in section 1.4 we will present major developments which cause these changes and in section 1.5 we will consider what makes these changes problematic and the different reactions to these problematic changes.

1.3 Types of environmental problems

Environmental problems manifest themselves as changes in the physical environment. Many human-induced changes in the physical environment are being interpreted as environmental problems. It is quite common to distinguish three broad categories of environmental problems: exhaustion, pollution and disturbance.

1.3.1 Exhaustion

The concept of exhaustion refers to the depletion of both non-living (abiotic) and living (biotic) nature. Natural resources are divided in non-renewable resources and renewable resources.

Non-renewable resources

Minerals and fossil fuels are non-renewable resources. Their stocks are fixed.

However, assessments of the quantities of available minerals and the moment of their exhaustion become outdated time and again. The precarious nature of such assessments is caused by the numerous uncertainties regarding the extractability of minerals and fossil fuels. Which quantities of materials are defined as available reserves depends on the contemporary state of geological knowledge, mining technology and state of the market. There are identified resources, hypothetical and speculative resources (which have not yet been discovered) and unconceived resources. Only part of the identified resources can be exploited in a way that is economically feasible.
The earth's crust also contains minerals which probably will never become available in substantial quantities because of the financial costs of extraction. The concept of 'potential economic threshold' refers to this barrier. Even harder to cross is the so-called 'mineralogical threshold'. Many minerals are present in large quantities, but in such low concentrations that extraction on a large scale must be considered impossible, because of the quantities of energy required. Another impediment is formed by the huge quantities of waste that will remain as an unwanted by-product of the extraction of these minerals.
The dimensions of the availability of resources categorized above are summarized in figure 1.2.

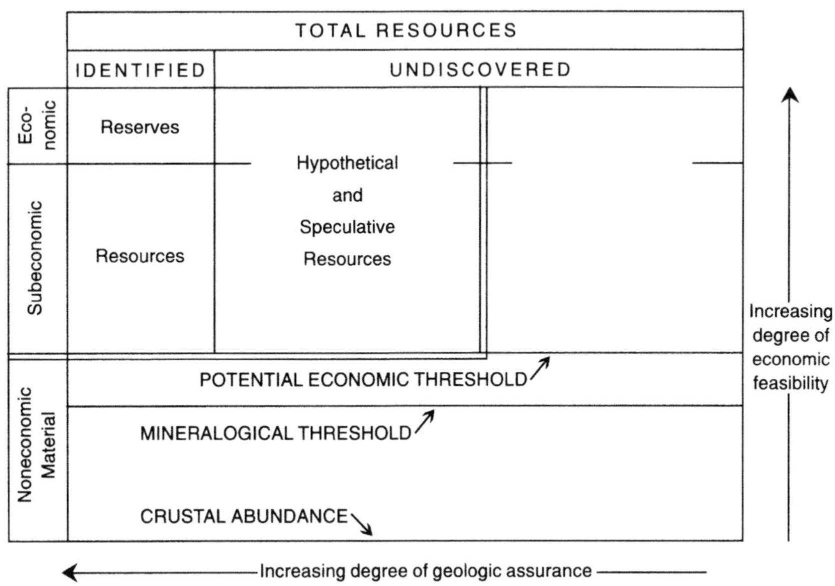

Figure 1.2 Classification of mineral resources.
Source: Brobst, 1979, p.118. Reprinted by permission of The John Hopkins University Press, © 1979.

So far social factors which limit the availability of minerals and fossil fuels have not been mentioned. For political reasons countries sometimes limit the extraction and the export to other countries. Extraction may also be limited because of harmful environmental effects. Different forms of pollution of air, water and soil, disturbances of ecosystems and damage to the scenic beauty of landscapes are among the negative environmental impacts associated with extraction of minerals and fossil fuels.

Renewable resources

The use of non-renewable materials often leads to their dispersal in air, in water or on soils. In that case, they are no longer available for human use. There are other elements of nature which can be consumed without excluding availability in the future. This is the case with the so-called renewable resources. These include both abiotic components like soil and water and biotic components like plants and animals. Their renewal can take place as a completely natural process or be stimulated by human intervention. Both rainfall and artificial infiltration can preserve stocks of groundwater when these stocks are used as drinking water. Even when plants and animals are used for human consumption their species can survive as a result of natural reproduction. The same effect can also be reached by deliberate breeding of plants and animals.

Exhaustion of renewable resources may take place directly or indirectly. Direct exhaustion takes place when the use of a resource exceeds its reproduction. This is the case when more groundwater is extracted than produced by rainfall or infiltration, when more fish are caught than reproduced, or more trees are felled than planted. Indirect exhaustion occurs when the conditions for reproduction are disturbed or when existing stocks are destroyed or made unusable. For instance, it is water pollution instead of hunting that limits the number of seals in the Waddensea along the coasts of the Netherlands, Germany and Denmark. At several places in the eastern part of the Netherlands, it was not the excessive consumption of water, but pollution by agricultural chemicals that made the further use of groundwater for the preparation of drinking water impossible. In theory, the use of renewable resources can be continued endlessly. However, this is only possible if stocks are not destroyed by external influences and if they are renewed permanently. In many cases of declining populations of plants and animals these conditions are not fulfilled. Various types of human use of natural resources lead to the shrinking of populations and the ultimate extinction of species of plants and animals.

Table 1.1 Number of recorded species extinctions since 1600

Group	Number extinct since 1600
Molluscs	191
Birds	115
Mammals	58
Other animals	120
Total animals	484
Plants	654

Source: UNEP, 1995 p.233

Renewable energy

Energy generation can be based on non-renewable or on renewable resources. Both uranium (fissile fuel for nuclear energy) and fossil fuels are non-renewable resources. The formation of natural gas, oil and coal takes so much time that, in studying contemporary environmental problems, these resources are considered as being non-renewable. Energy can be provided by means of sunshine, wind and water power. These types of energy do not use stocks and for that reason cannot be exhausted. In particular for the third world countries, wood and other forms of bio-mass are important renewable sources of energy.

In theory, the use of these 'endless' sources of energy is the least harmful to physical environments. In practice, even the use of these resources is causing environmental problems. In countries like Brazil and Russia, production of hydroelectric power led to the loss of extensive areas, their flora and fauna, their settlements and their agricultural land. Even the construction and the use of windturbines for electricity production may cause damaging environmental impact, for example on birds.

1.3.2 Pollution

Used materials, whether or not transformed into solid, fluid or gaseous waste, sooner or later return into the environment.

In case of exhaustion the environment is used as a storehouse which is depleted. In case of pollution the environment functions as a garbage can which overflows. Pollution can be defined as the release of harmful substances into the environment. Besides pollution, as defined above, there are other environmental effects of human behaviour which are experienced as annoying or risky. Examples are: noise, radiation and heat. The term 'environmental stress' is used as the all-embracing designation of these various effects. In chapter 7 on risk and nuisance we will discuss human perceptions and responses to different forms of environmental stress. Here we deal only with different forms of pollution.

In the preceding section we made a distinction between renewable and non-renewable resources. Here we will distinguish between degradable and non-degradable or persistent substances.

Degradable substances

Emission of substances into air, water and soil is a permanent and natural process. Waste substances are used as nutrients by bacteria's. These organisms transform substances in such a way that they can be used again as nutrients by plants and, as a consequence, also by animals. Millions of bacteria's and other organisms may live in one gram of soil. These *decomposers* prepare the remnants of plants and animals for absorption by new plants and new animals. (see figure 1.3)

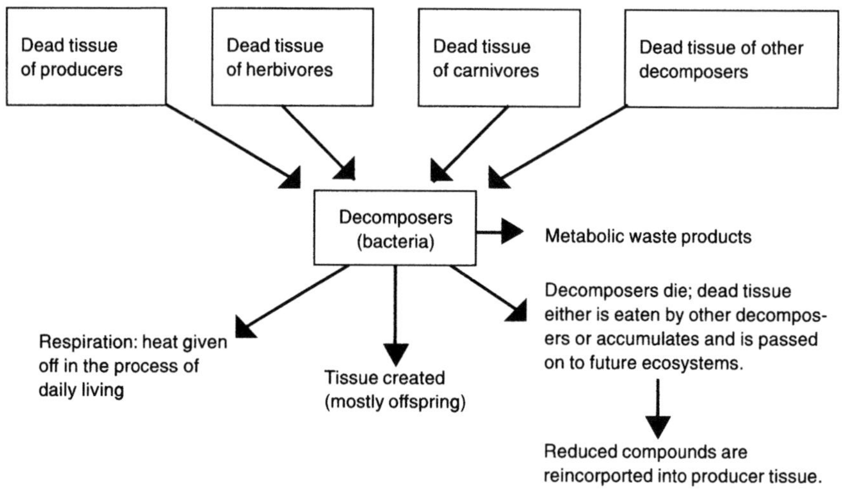

Figure 1.3 Inputs and outputs of a decomposer
Source: Sutton and Harmon (1973, p.52) Reprinted by permission. © Copyright John Wiley & Sons, Inc.

Degradation and mineralization of waste is a process as natural as the renewal of resources discussed earlier. In both cases human interventions can stimulate natural processes. People can make compost out of their own vegetable, fruit and garden waste. In agriculture, animal and human faeces can be used to enhance soil fertility. Even the environmental problems related to the renewal of resources and degradation of substances resemble each other. Both the use of renewable resources and landfilling with waste can exceed the capacities of the environment, to renew resources and to degrade substances respectively. Manure surpluses caused by intensive cattle breeding are an example of overburdening the natural capacity to degrade substances.

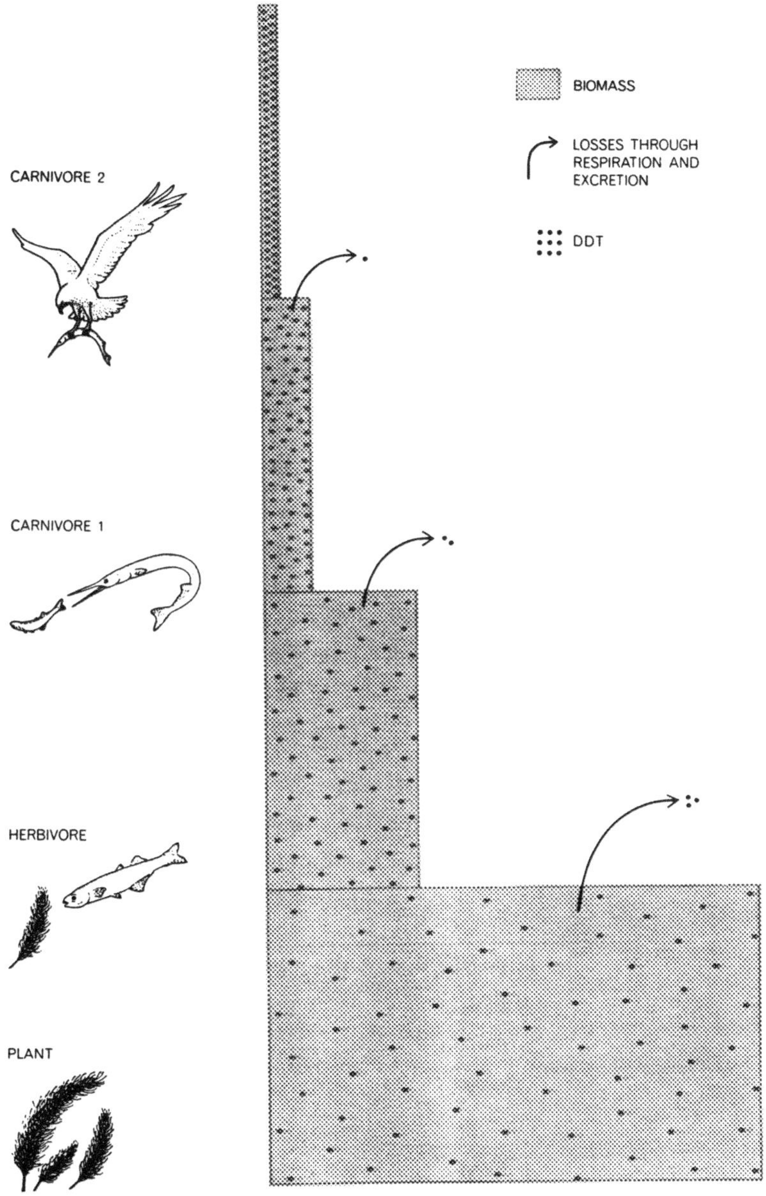

CONCENTRATION of DDT residues being passed along a simple food chain is indicated schematically in this diagram. As "biomass," or living material, is transferred from one link to another along such a chain, usually more than half of it is consumed in respiration or is excreted *(arrows);* the remainder forms new biomass. The losses of DDT residues along the chain, on the other hand, are small in proportion to the amount that is transferred from one link to the next. For this reason high concentrations occur in the carnivores.

Figure 1.4. Accumulation in the food chain.
Source: Woodwell (1967, p.30)

Non-degradable or persistent substances

Nowadays many substances are produced which do not exist in nature. Some of these substances cannot be degraded by natural processes at all or only very slowly. Among these man-made substances are pesticides which are used in agriculture. They can be divided into insecticides, herbicides and fungicides which are used to destroy insects, weeds and fungi respectively. One of the most infamous insecticides is DDT. In 1962 Rachel Carson described in her book *'Silent Spring'* the death of birds caused by the use of DDT and other pesticides. Probably no other book contributed so much to the raising of awareness of environmental problems during the sixties (See section 6.5.2). Today the use of DDT is prohibited in a large number of countries. Nevertheless, like many other pesticides, DDT is still present in the environment today.

For a clear understanding of the effects of persistent substances like DDT the concepts of 'food chain' and 'food pyramid' are very important. The base of the food pyramid is formed by autotrophic organisms which prepare their own food out of energy and build themselves up by means of anorganic substances like water, carbon dioxide and salts. Plants belong to the autotrophic organisms. By means of absorption of sunlight the leaf green (chlorophyll) of plants is able to transform water and carbon dioxide into glucose, oxygen and water. This process is called photosynthesis. Heterotrophic organisms are dependent on other organisms for their food supply and feed on organic substances of plants and animals. At each following trophic level the quantity of organic matter or biomass is reduced by a factor of ten. Persistent substances like DDT are not reduced to the same degree. As a consequence, the concentration of these substances increases at each higher trophic level. This process is called accumulation in the food chain. As a result of this process, substances which are present in the environment in low concentrations may end up in high concentrations in human food.

The phenomenon of reduction of biomass at each higher trophic level is sometimes used as an argument for vegetarianism or reduction of meat consumption. In using vegetable food of the lower trophic level of plants, less biomass is needed than in case of consumption of meat of the higher trophic level of animals.

Flux-type and sink-type pollution

Forms of environmental pollution can be divided into flux-type and sink-type pollution. Well-known forms of flux-type pollution are air pollution and surface water pollution. In cases of flux-type pollution we are dealing with streams of pollutants. This type of pollution can only be reduced by reducing the emission of these pollutants. Forms of pollution which are now called flux-type were already perceived as environmental problems thousands of years ago. Environmental hygiene measures as developed in western industrial states in the nineteenth century were aimed at the reduction of concentrations of pollutants in air and water. In recent decades several countries have been successful in reducing the concentrations of sulphur dioxide in the air and of heavy metals in surface waters. The reduction of concentrations of heavy metals in the water of the river Rhine is an example of the latter development. As far as the concentration of some major polluting substances is concerned, air and water have become cleaner at many places in recent years.

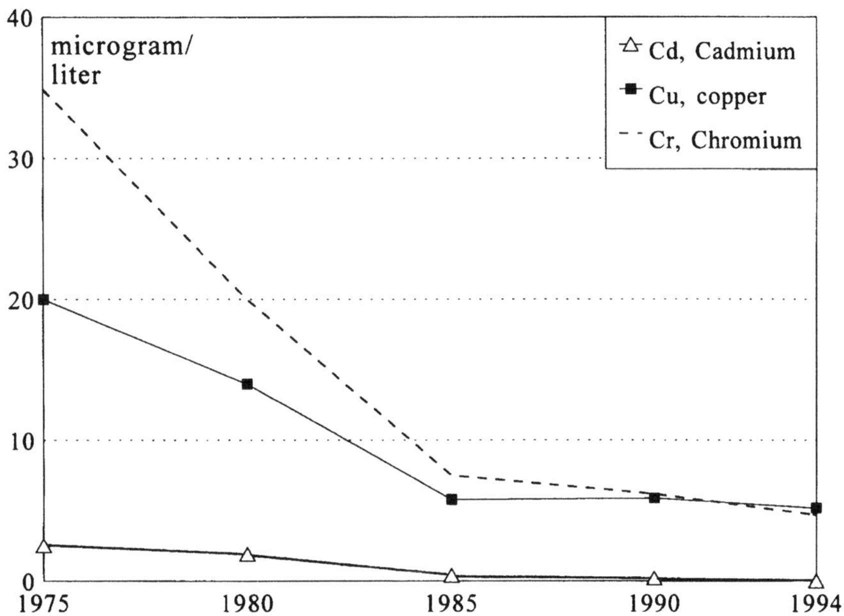

Figure 1.5 Reduction of heavy metals in the water of the river Rhine at the German-Dutch border
Source: Netherlands Central Bureau of Statistics

Reduction of concentrations of substances in air and water does not end the accumulation of these substances at fixed places like soils and river beds. This process of accumulation is called here sink-type pollution. Several contemporary environmental problems such as the pollution of soils, river beds and groundwater wells as well as acidification and the enhanced greenhouse effect are cases of sink-type pollution. Lowering the concentrations of heavy metals in river water does not stop their accumulation in river beds, but only reduces the speed of this process. The same can be said of the lowering of the concentration of acidifying particles in the air. It will retard but not necessarily stop the accumulation of acid particles in woods and as a consequence the forests may continue to die as well.

1.3.3 Environmental disturbance

The concepts of exhaustion and pollution were elucidated with the images of the storehouse and the garbage can respectively. Not all environmental problems are covered by these images. In the environment there are complex relationships between biotic and abiotic elements. Environmental problems do not only originate in taking too much out of the environment (exhaustion) or dumping too much in it (pollution) but also in the disruption by human activities of relationships which occur and processes which take place between different elements of the environment. The concept of disturbance is used here to refer to such types of environmental problems. There is a great variety of environmental relationships and processes which can be disturbed. Both the relationships between elements of a natural area which are highly valued by biologists and those of a landscape that is appreciated by, for example, painters can be disturbed. The totality of interdependent abiotic and biotic elements in a certain area during a certain period is called an *ecosystem.* Environmental problems are often related to the complicated relationships between abiotic and biotic elements of nature. In addition to the examples of relationships which were described in section 1.3.1. we give two more examples:

- Emissions of waste substances in surface waters may stimulate the growth of plants. As a consequence of the oxygen use of these plants the concentration of oxygen may drop so far as to make animal life impossible. In the case of this phenomenon, which is known as water bloom or eutrophication, environmental elements are not directly damaged. On the contrary, the direct effect is algal growth. However, the indirect effect is stinking water and starvation of fish.
- An apparently innocent interference like the cutting down of hedges to facilitate large scale agricultural production may lead to the disappearance of birds and, as a consequence, to the increase of crop-damaging insects. This may encourage the use of pesticides. An alternative to the use of dangerous chemicals is offered by biological or integrated control. This method stimulates the development of natural conditions which reduce elements like insects and weeds which hinder agricultural production.

Environmental problems are sometimes defined as disturbances of environmental functions. As long as natural processes are not disrupted by human activities, nature will fulfil all sorts of useful functions such as productive functions (the provision of vegetable and animal food) and cleaning functions (the conversion of waste materials into useful substances). These and other useful environmental functions may be disturbed by human activities and therefore necessitate other human activities to replace the disrupted spontaneous natural processes.

Table 1.2 Examples of functions of ecosystems

service/good of ecosystems	**ecosystem function**
Gas regulation	Regulation of atmospheric chemical composition
Climate regulation	Regulation of global temperature, precipitation, and other biologically mediated climatic processes
Disturbance regulation	Capacitance, damping and integrity of ecosystem response to environmental fluctuations
Water supply	Storage and retention of water
Erosion control	Retention of soil within an ecosystem sediment retention
Nutrient cycling	Storage, internal cycling, processing and acquisition of nutrients
Waste treatment	Recovery of mobile nutrients: removal, breakdown of excess of xenic nutrients and compounds
Refugia	Habitat for resident and transient populations
Food production	Portion of gross primary production extractable as food
Raw materials	Portion of gross primary production extractable as raw materials
Genetic resources	Sources of unique biological materials and products
Recreation	Providing opportunities for recreational activities
Cultural	Providing opportunities for non-commercial activities

From a list of 17 services and functions by Costanza *et al.* (1997, p.254). Reprinted by permission.

Not all environmental problems manifest themselves as directly visible and comprehensible phenomena like the cutting down of a beautiful tree, the dumping of oil at sea or the exhaustion of a stock of fossil fuels. Often there are complicated relationships involved which can only be discovered by means of scientific research.

1.4 Causes of environmental problems

It is impossible to prevent all environmental problems. Human beings cannot live without permanently changing their environment. Already by their metabolism, by their breathing and defecation, people change their environment. Human beings have to change their environment to protect themselves against climatic influences like high or low temperatures, precipitation, sunshine, wind and floods and against poisonous plants and dangerous animals. They have also to provide themselves with water, food, materials and heat. Many human activities are directed towards the protection against undesirable natural influences and obtaining of desired natural products. It is unavoidable that these types of human changes in the environment are sometimes interpreted as problematic either by the people who caused these changes themselves or by others. This phenomenon is inherent to human existence.

However, the growing impact of human behaviour on the physical environment in recent decades has brought environmental problems on to political agendas all over the world. Nowadays there are three main factors which contribute to that growing impact: increasing numbers of people, and the growing quantity and changing quality of human use of the environment. Below we will discuss these three causes of environmental problems.

1.4.1 Population growth

The more people are living together, either locally or world-wide, the more likely human influence on the environment will be experienced as problematic.

Already one and a half century ago the English priest and economist Robert Malthus (1766-1834) was convinced of the limited capacities of the earth to feed a growing world population.

Malthus on limited capacity

The main peculiarity which distinguishes man from other animals, in the means of his support, is the power which he possesses of very greatly increasing these means. But this power is obviously limited by the scarcity of land – by the great natural barrenness of a very large part of the surface of the earth – and by the decreasing proportion of produce which must necessarily be obtained from the continual additions of capital applied to land already in cultivation.

Malthus, 1976, p.225.

According to Malthus, human beings, like other animal species, are able to reproduce geometrically (1, 2, 4, 8, 16 etc) whereas food production increases, at most, arithmetically (1, 2, 3, 4, 5 etc). Population growth will be limited either by 'preventive' or by 'positive' checks. Preventive checks include all forms of limitation of child birth. Positive checks are defined as 'all the causes which tend in any way prematurely to shorten the duration of human life' (Malthus,

1976, p.250). The message of Malthus was that if child birth is not restricted, population growth will be limited by lack of food or other forms of suffering. Social legislation and other measures to protect poor people will not help them to escape from the misery caused by positive checks in the case of unrestricted population growth.

Karl Marx (1818-1883) rejected the idea that famine was caused by over-population. He illustrated his argument with the Irish famine of 1846 that killed more than a million of people but 'did not do the slightest damage to the wealth of the country'. (Marx, 1976, p.862). According to him, not the sheer numbers of people, but the developments in agriculture and industry, caused the misery in Ireland.

Capitalism and surplus population in 19th century Ireland

'The smaller and the medium farmers – I reckon among these all who do not cultivate more than 100 acres - still make up about 8/10 of the whole number. They are, one after the other, and with a degree of force unknown before, crushed by the competition of an agriculture managed by capital, and thus they continually furnish new recruits to the class of wage-labourers. The one great industry of Ireland, the manufacture of linen, requires relatively few adult men, and only employs altogether, in spite of its expansion since the price of cotton increased in the years from 1861 to 1866, a comparatively insignificant portion of the population. Like all other large-scale industries, it constantly produces, owing to its incessant fluctuations, a relative surplus population within its own sphere, despite the absolute increase in the mass of human beings absorbed by it'

Marx, 1976, pp.862-863.

The controversy between Malthus and Marx resembles contemporary discussions about the role of population growth as a major cause of environmental problems. In the seventies Paul and Anne Ehrlich (1970) emphasized the role of population growth as the main cause of the contemporary environmental crisis, whereas Barry Commoner (1971) ascribed only a minor role to population growth as a factor contributing to environmental stress. The statistics used by Commoner referred to the United States. In recent decades, in wealthy western countries all kinds of environmentally damaging activities have increased much more than the sheer numbers of human beings.

So it can be concluded that population growth is not the only major cause of increasing environmental stress.

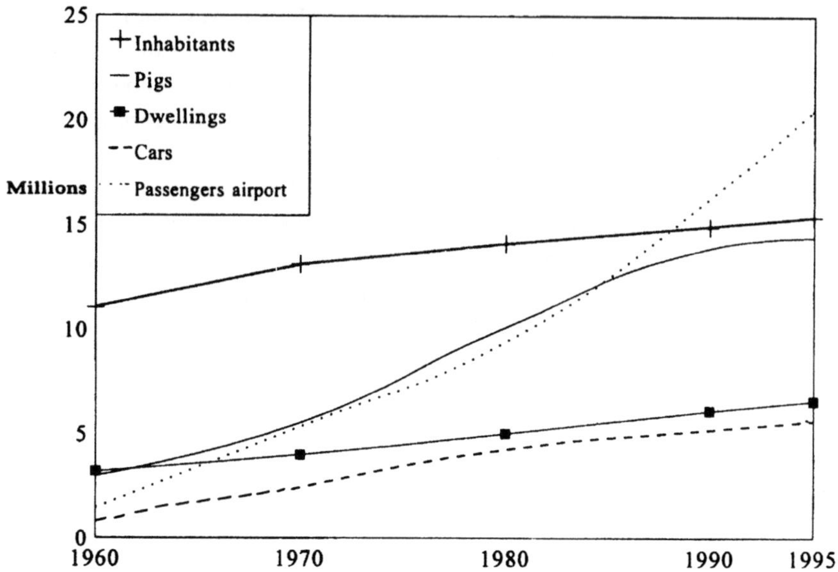

Figure 1.6. Growth of population compared with growth of other environmental stress factors in the Netherlands.
Source: Netherlands Central Bureau of Statistics.

1.4.2 Quantity of environmental utilization

Not only the growing numbers of people but also the growth of their *per capita* use of environmental resources contributes to contemporary environmental problems. Practically all countries of the world consider economic growth, expressed in growth of the Gross National Product (GNP), as one of the most, if not the most, important policy goals. In general, economic growth leads to a larger utilisation of natural resources. There is often a strong correlation between economic growth on the one hand and growth of energy use on the other hand. For that reason economic growth, together with population growth, has often been considered as a major cause of environmental problems. A disadvantage of using the term 'economic growth' in this context is that it is strongly associated with private production and consumption. However, public activities like the construction and use of military training grounds and warfare can also cause environmental problems. The same can be said of all sorts of unpaid human activities, like motorized recreation, which are not included in figures for economic growth but which do cause environmental problems. As indicators of the size of environmental stress data about economic growth are of limited value.

1.4.3 Quality of environmental utilization

Beside the neglect of public and unpaid private activities there is still another limitation of the usefulness of economic growth as an indicator of environmental utilization. The production of a fixed amount of GNP may be associated with widely different quantities of materials used and pollutants emitted. In the early seventies Barry Commoner showed that since 1946 the provision of basic goods like food, clothing and shelter in the United States grew in proportion to the growth of population. However, according to Commoner, the environmental impact of the production of these goods increased in that period. Commoner used the term 'technology' to indicate this qualitative change of production with damaging environmental consequences.

Bad technology as cause of environmental problems

'What emerges from all these data is striking evidence that while production for most basic needs – food, clothing, housing – has just about kept up with the 40 to 50 per cent or so increase in population (that is, production *per capita* has been essentially constant), the *kind* of goods produced to meet these needs have changed drastically. New production technologies have displaced old ones. Soap powder has been displaced by synthetic detergents; natural fibres (cotton and wool) have been displaced by synthetic ones; steel and lumber have been displaced by aluminium, plastics, and concrete; railroad freight has been displaced by truck freight; returnable bottles have been have been displaced by nonreturnable ones. On the road, the low-powered automobile engines of the 1920's and 1930's have been displaced by high-powered ones. On the farm, while per capita production has remained about constant, the amount of harvested acreage has decreased; in effect, fertilizer has displaced land'.

Commoner, 1971, pp.141-142.

'Technology' is not the right term to refer to the qualitative aspect of human influences on the physical environment. This qualitative aspect is not only dependent on technological developments but also on organizational or managerial developments. Whether production processes or products are more or less exhausting or polluting the environment is not only determined by technology but also by the way industrial production is organized. Unlike Commoner we prefer the phrase 'quality of environmental utilisation' rather than 'technology' as an indication of the qualitative aspect of human impact on the physical environment. Developments in recent years make remarks like those of Commoner about the impact of technology one-sided and therefore outdated. The quality of human-induced change of the environment does not always develop in a negative way. In the past twenty years modernisation in industrial production processes and of industrial products has resulted in a reduction of the negative environmental effects of these processes and products.

The term 'ecological modernization' has been introduced to indicate this development (Huber, 1982). In principle it is possible to reduce the environmental impact of a given level of consumption or life style by changing production processes and the composition of products. So changing the quality of environmental utilization can both add to the combined effect of growth of population and production and reduce this combined effect. The first process was described by Commoner, the second process by authors recently discussing 'ecological modernization'.

1.4.4 Carrying capacity

The possibilities for human use of natural resources are limited. The concept of *carrying capacity* refers to the limits of human activities that physical environments, either locally or world-wide, can endure. However, carrying capacity is not a static but a dynamic phenomenon. Both the availability of natural resources and nature's capacity for waste disposal can be enlarged.

The history of agriculture and the different ways in which it is practiced today offer many examples of augmentation of carrying capacity. In the course of time, yields per hectare and per worker sharply increased. As a consequence, in theory it would be possible to feed numbers of human beings several times larger than the present world population. Nevertheless, agriculture also shows what can go wrong in trying to enlarge the earth's carrying capacity. Damaging effects of intensification of agriculture often manifest themselves at other places and at later periods. Chemicals which are used in agriculture are transported by air and water to other places and may have damaging effects there. Future generations may suffer from salinization or soil erosion caused by intensive agriculture in the past.

Carrying capacity is not only a dynamic phenomenon because it is dependent on the state of technology but also because it is dependent on changing cultural values. It is not only determined by what the earth can deliver and dispose of, but also by what people can and are prepared to endure.

1.5 Interests, values and reactions to environmental problems

We have based our categories of environmental problems on their different objective physical manifestations. However, as we argued before (see section 1.2), human induced environmental changes only become environmental problems if these changes are considered problematic by human beings. There may be great differences of opinion about the problematic character of human-induced changes of the environment.

1.5.1 Interests

One way of interpreting these differences is to consider them as manifestations of conflicting interests. A farmer and a conservationist often have different opinions about the most desirable development of the countryside, because they have different interests. The former wishes to transform nature in order to achieve the highest yields with the least effort. The latter wants to protect existing natural conditions in order to conserve biodiversity and scenic beauty. Human-induced changes in the environment that may be beneficial to agricultural production, transport or housing, may be damaging to the protection of natural beauty or to recreation. In the following chapters we will discuss different examples of conflicts of interests about desirable environmental qualities.

1.5.2 Values

Another way of considering the various opinions regarding environmental problems is to relate the differences to human values. Whether human-induced changes in the environment are considered problematic depends on the prominence of widely different values. Some of these values are related to the importance attached to undisturbed nature. Examples are: respect for undisturbed nature as such, appreciation of beauty and rarity of nature and of the scientific importance of elements of nature. Other relevant values are related to the environmental effects of human activities. Among them are religious taboos and moderation in production and/or consumption. A third category of relevant values concerns individual and social well-being. Relevant values for this category are (among others): health, welfare, survival of the human race and equitable division of material goods. In chapters 5 and 6 we will discuss more extensively the importance of values for the understanding of environmental problems.

1.5.3 Reactions

There is no one to one relationship between human interests or values on the one hand and human behaviour on the other hand. A major aim of this book is to understand the effects on human behaviour of those human-induced environmental changes which are in conflict with human interests and values. Why, when and how do people react to environmental changes which are in conflict with their interests and values? What is the relation between environmental attitudes and environmental behaviour, which types of environmental risks and nuisance are reacted upon? In what way and when do people react to environmental problems by creating new, or supporting existing, environmental organizations and how do these organizations operate? What factors are decisive in determining whether or not environmental problems are reaching political agendas and whether policy measures are successfully put in practice? The discussion of these types of questions may be considered the main theme of this book. Before discussing them, in the next chapters we will examine the historical, geographical and cultural context of environmental problems.

Chapter 2

History

2.1 A problem down to all ages?

The 'environment' is an issue of the last thirty years. But even in the remote past human beings changed their environment in a way which was experienced as problematic either by themselves, their contemporaries or by later generations.

Environmental issues 2000 years ago

In the earth itself you often see sulphur generated and malodorous asphalt congealing. When men are following veins of gold and silver, groping with their picks in the bowels of the earth, what fumes are emitted from the pits of Scapte Hyle! What malignant breath is exhaled by gold mines! How it acts upon men's features and complexions! Have you not seen or heard how speedily men die and how their vital forces fail when they are driven by dire necessity to endure such work? All these vapours, then, are given off by the earth and blown out into the open, into the unconfined spaces of the air.

Lucretius (c.100 - 55 B.C.), 1951, p.242.

So there is nothing new under the sun? On the contrary, major contemporary environmental problems like the enhanced greenhouse effect, the depletion of the ozone layer, the disappearance of tropical rainforests and the extinction of plants and animals are caused by a process of radical changes in society during the last centuries. This process is often called 'modernization'. It started in Europe but has spread all over the world.

Classical sociologists like Ferdinand Tönnies (1855-1936), Emile Durkheim (1858-1917) and Max Weber (1864-1920) contrasted modern society, as it first developed in Europe, with all preceding types of societies. Tönnies (1887) described the transition from 'community' to 'society'. In this process former communal bonds disappeared and were replaced by non-personal, commercial bonds. Durkheim (1993) argued that social bonds based on common convictions were weakened and concluded that in modern societies their role as a major binding force should be taken over by the division of labour. Max Weber (1968) was fascinated by the unavoidable process of rationalization that penetrated all

spheres of life.
The most important environmental aspects of the process of modernization are extreme population growth and fundamental changes in quantity and quality of the human use of natural resources. We will illustrate these changes by comparing modern societies with earlier types of societies, followed by a discussion of the limits of natural resources to human exploitation.

2.2 Evolution of modes of production

In chapter 1.2, a distinction was made between natural and environmental catastrophes. We cannot characterize the historical development of relations between men and their environments better than by saying that human beings are less and less threatened by natural catastrophes and more and more by environmental catastrophes. The all powerful and threatening nature comes more and more under human control, but the consequences of human behaviour on nature cause more and more problems and increased risks.
The carrying capacity of nature (chapter 1, section 1.4.4) always sets limits to the human use of natural resources. Societies may react upon these limits in two different ways. They can accept existing limits and restrict their use of natural resources, or they can, by means of technical and organizational measures, try to increase the carrying capacity of the environment.
From studies of societies in ancient history we can learn that even then restrictions of the human use of natural environments already existed. Even in early history, when the number of human beings on earth was small, the consumption *per capita* was modest and polluting factories did not yet exist, measures were taken which kept the human use of natural resources within the boundaries of the carrying capacity of their environment. 'Successful adjustments in the past have involved population controls, expansion of living area achieved either through migration to a new frontier or annexation of the territory of others, or new technological capabilities' (Bilskey *et al.*, 1980, p.7).
It is a myth that in ancient times population growth was only restricted by mortality and other natural circumstances. Social factors like *post-partum* abstinence, ceremonial abstinence associated with religious feasts or rituals, or during periods of mourning, *coitus interruptus*, abortion and infanticide contributed to the limitation of population size (Benedict, 1970, pp.166-175)
Whether and to what extent such measures were taken with the intention to protect the environment is not clear. However, the result was that stress on the local environment was limited.
As long as – according to the bible – the earth was still 'formless and empty', environmental degradation as a result of excessive exploitation could be prevented by migration to other territories. In Genesis 13 a situation is described in which the problem of scarcity of natural resources was solved by migration.

Abraham and Lot possessed herds which together exceeded the carrying capacity of their grounds, resulting in quarrels between their herdsmen. Abraham proposed to divide available grounds and offered Lot the choice of the best region. Things were not always arranged quite so peacefully. The acquisition of natural resources has always been one of the most important driving forces for warfare and conquest of territories. Both the colonization of large parts of the world by European countries and the more recent war in Kuwait were inspired by the need for natural resources.

Another way to escape from scarcity of resources is to shift to another way of exploitation of the environment. Different types of society can be distinguished by the way in which they use their environment. In *hunting and gathering societies* men live from vegetables and animal foods which are available in nature. In *agrarian societies* soils are systematically treated and plants are cultivated and animals are raised for human consumption. In *industrial societies* minerals are extracted and used as raw materials and sources of energy and processed more and more to satisfy human needs.

2.2.1 Hunting and gathering societies

The combination of gathering, hunting and fishing has been described by Roberts as a mode of production that generally uses and manipulates natural ecosystems without changing them (Roberts, 1989, p.122). But even this author mentions the extinction of great mammals on the American continent between 14,000 and 10,000 years ago. It is assumed that hunters/gatherers contributed both directly (by hunting) and indirectly (by destruction of habitats) to this extinction. In north America, in a period of only a few thousands of years, 39 great species of mammals died out. This is more than in the preceding four million years (Roberts, 1989, p.61).

Despite the modest development of their technical means hunting and gathering societies adapted successfully to their natural environment. The efforts hunters and gatherers had to make in order to provide their vital necessities were far less than those of farmers in later agrarian societies. In this context the concept of 'the original affluent society' has been used (Sahlins, 1972). With a working day of three to five hours people were able to provide their vital necessities. All the resources that were needed were free available in their natural environment. Welfare was not based on a high level of production but on a low level of consumption. For the mobile hunter/gatherer possessions were a burden that had to be limited. Together with a low consumption level and a great mobility, limitation of population size was necessary to survive as hunter/gatherers. Drastic measures like killing babies and old people sometimes were not avoided in the struggle for survival.

Equality among productive people

The distributive process in hunting and gathering societies is radically different from that in industrial or agrarian societies. There is comparatively great equality in the division of the products of hunting and gathering. The results of these efforts are necessary to maintain the productive power of the members of society. Unlike the situation in later agrarian societies, there is no economic surplus which can be appropriated by a powerful elite. However, persons who are not able to contribute to productive activities are highly vulnerable. Babies and old people as well as sick and injured may be eliminated by abandoning them.

Lenski, 1966, pp.102-104

2.2.2 Agrarian societies

The failure of existing forms of environmental exploitation to provide vital necessities stimulates the shift to alternative ways of exploitation (Wilkinson. 1973). Such a decline can be caused both by changing natural conditions, like climate change, and by human-induced environmental change. As a consequence of population growth the yield *per capita* of the existing form of environmental exploitation fall short (Boserup, 1965). Between 15,000 and 10,000 years ago the 'neolithic revolution' took place: the shift from gathering and hunting to agriculture and cattle rearing.

Agriculture can be practiced in widely different ways. In the case of shifting cultivation soil is prepared for agriculture by means of slashing and burning. Ashes resulting from burning are used to enrich the soil. As soon as soil becomes exhausted, new areas are taken into cultivation. After periods of laying fallow soils can be used again. 'This strategy required elaborate technical and social skills. It presupposed not only the availability of strong and sharp axes but also the capacity to plan ahead for at least several months.' (Goudsblom, 1994. p.49.)

The opposite of shifting cultivation, with only temporary use of soil, is permanent farming. By shortening the period of fallow, better soil cultivation, manuring and weeding become more important. In dry areas irrigation can be put into practice. In such ways the carrying capacity of an area can be increased. This does, however, require more human effort. Typical features of extensive shifting cultivation are: low labour productivity, high yields per labour unit and low yields per surface-unit. High labour intensity, high yields per surface-unit, and low yields per labour-unit are typical of intensive permanent farming (Kloos, 1978, p.50).

Compared with hunting and gathering societies, agrarian societies are characterized by greater social inequality. In agrarian societies both wealthy elites and groups of people suffering almost permanently from starvation exist next to each other.

Who profits from surplus?

In any society, wealth implies that a surplus has been wrung from nature, that society has not only solved its economic production problem but has achieved a margin of effort above whatever is required for its own existence. Perhaps what first astonishes us when we regard the civilizations of the ancient world is the size of surplus that could be got from basically poor peasant populations. The temples of the ancient Assyrian kings, the extraordinary treasures of the Aztecs, the pyramids and pleasure craft of the pharaohs of Egypt, the Acropolis of Athens, and the magnificent roads and architecture of Rome all testify to the ability of an essentially agricultural civilization to achieve a massive surplus, to pry considerable amounts of labour loose from the land, support it at whatever low level necessary, and put it to work building for posterity(....). To whom did the wealth of antiquity accrue? (......). Most wealth did not go to those who played a strictly economic role(....). Rather, in ancient civilization, wealth was generally the reward for political, military, or religious power or status, and not for economic activity.

Heilbroner, 1993, p.23.

There are different types of environmental damage caused by agricultural societies in ancient times. One of them is soil erosion as a consequence of deforestation. As early as 6000 BC, in central Jordan soil erosion caused by deforestation led to declining crop yields and forced people to abandon their villages (Ponting, 1992, p.69). Another example of environmental stress in agricultural societies is salinization caused by irrigation.

2.2.3 Intermezzo: irrigation and power

Agrarian societies also developed in areas with difficult natural circumstances and where drought hampered agriculture. Agriculture can be made possible in dry areas by means of irrigation. Irrigation can be developed on a small scale by single farmers or agricultural communities. However, it can also be implemented on a large scale by means of a complex system of watermovement across long distances. This was already happening thousands of years ago in several parts of the world.

There are well-known historical examples of environmental decay as a consequence of intensive agriculture. The decline of Mesopotamia has been attributed to salinization caused by irrigation. Between 2400 and 1700 B.C. as a consequence of salinization, corn harvests went down from 2530 litre to 897 litre per hectare and the region lost its position as a political and cultural centre (Jacobson and Adams, 1958).

The sociologist Max Weber argued that in ancient times in Egypt, Mesopotamia and China the construction and maintenance of large irrigation works was related to the development of a centralized authority and bureaucracy. Later Karl Wittfogel (1957) introduced the concept of 'hydraulic society' to describe a type of society in which the maintenance of a large scale system of irrigation led to the development of central authority with despotic power. The subordinated population was not only obliged, whether or not by statutory labour, to contribute to the maintenance of the irrigation system, but also had to reward the authorities in money or in kind. This enabled the authorities to live in wealth and to maintain a military power.

With the development of industrial societies the phenomenon of societal power based on the development and maintenance of irrigation systems, as described by Weber and Wittfogel, did not come to an end. In his 'Rivers of empire' Donald Worster (1985) described - referring to the work of Wittfogel - how in the past one and a half centuries in the west of the United States irrigation led to an ever growing concentration of power, of which powerless citizens and the environment became victims. Worster describes the development as a process with its own dynamic in which thirst for power was often larger than thirst for water, which explains why more expensive measures to enlarge the supply of water were preferred above cheaper measures to reduce wastage of water.

In the context of the developments in the west of the United States Worster uses the concept of 'hydraulic trap'. This concept also applies to the development of irrigation in the former Soviet Union. In several republics of the former USSR large scale water management projects were undertaken for hydroelectric power and irrigation. Human settlements and thousands of acres of fertile soil were sacrificed to the creation of large reservoirs for hydroelectric power plants. Excessive irrigation destroyed fertile soils, which could no longer be used for food production. The area of salt-contaminated agricultural soils was estimated at 157 million hectares (Yablokov, 1990, p.4). What has been written about the causes of excessive and damaging irrigation in the Soviet Union looks similar to Worster's description of the situation in the United States. The common element in both descriptions is the rise in power and independence of a state office which has a vested interest in the enlargement of its services and is not motivated towards reduction of water consumption by conservation measures.

2.2.4 *Industrial societies*

Traditionally, the industrial revolution, the transition from agrarian to industrial societies, is narrowly fixed both in space and time: England in the second half of the eighteenth century. It is the period of the development of the steam engine and of several inventions for textile production. The introduction of the steam engine was preceded by a long process of gradual replacement of wood by coal as the main fuel. Coal had already been used for centuries. In the sixteenth and seventeenth century replacement of wood by coal was stimulated by rising wood prices as a consequence of increasing scarcity of wood. The textile inventions enabled the rapid development of the cotton industry. The development of cotton industry in England was closely connected with the development of Great Britain as a colonial world power. As a colonial power Britain was preceded by Portugal at the end of the fifteenth century. But Britain was the first country which used its colonies to stimulate domestic production. 'Between 1700 and 1750 home industries increased their output by seven per cent, export industries by seventy-six per cent; between 1750 and 1770 (which we may regard as the runway for the industrial 'take-off') by another seven per cent and eighty per cent respectively' (Hobsbawm, 1979, p.48). Since then, the international dependency of industrial countries on the import of raw materials and export of final products has increased.

The start in Britain

The British revolution was the first in history. This does not mean that it started from zero, or that earlier phases of rapid industrial and technological development cannot be found. Nevertheless, none of these initiated the characteristic modern phase of history, self-sustained economic growth by means of perpetual technological revolution and social transformation.

Hobsbawm, 1979, pp.34-35.

There are great differences between processes of industrialization in different periods and different countries. In recent processes of industrialization in Japan and the so-called new industrial states like South Korea and Taiwan, the role of the state is far more prominent than it was in Britain in the eighteenth century. The complex process which enables the development of industrial production is usually called *modernization*. In general, the implicit assumption of studies on industrialisation and modernization is that these are beneficial developments. Modernization is a must. Modernization implies striving for continued economic growth. It is often considered a universal process that sooner or later will take place all over the world. In a study that became a classic, Rostow even defined five universal stages of modernization:

- The traditional society
- The preconditions for take-off
- The take-off
- The drive to maturity
- The age of mass-consumption (Rostow, 1962, p.4.)

Not only this 'non-communist manifesto' but also the Communist Manifesto of Marx and Engels can be read as an ode to modernization, as it originated under capitalistic circumstances.

Capitalism as an economic growth machinery

Capitalism as a whole proved an unparalleled machine for the accumulation of capital. In its development we find the first economic system in history in which economic growth became an *integral* part of daily life. As Marx and Engels were to write in the Communist Manifesto: 'The bourgeoisie, during its scarce one hundred years, has created more massive and more colossal productive forces than have all preceding generations together.' And the compliment, all the more meaningful coming from the two archenemies of its social order, was true.

Heilbroner, 1993, pp.89-90.

2.3 Environment and modernization

In chapter 1.3 we mentioned the growth of population and the development of quantity and quality of environmental utilisation as causes of environmental problems. When the development of population growth and growth of environmental utilisation since the industrial revolution are shown graphically, steep curves are the result (See figure 2.1. and 2.2.).

2.3.1 Population growth

It took the human species hundreds of thousands of years to reach - around 1850 – the number of one billion people. Once the number of one billion was reached, the time span until another billion had been added to the world population, continued to decrease. This is shown in figure 2.1.

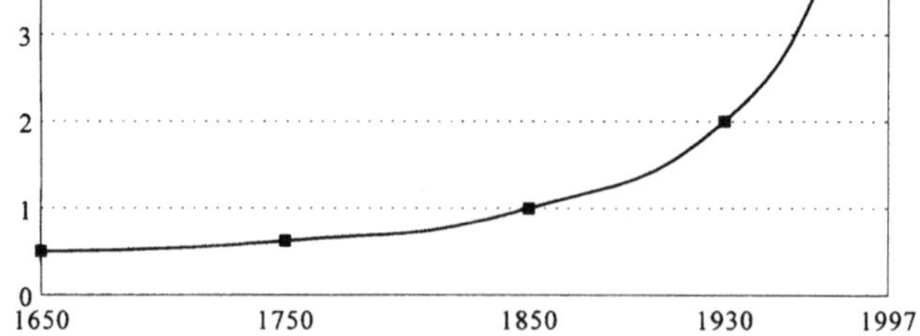

Figure 2.1 Growth of the world population

Population growth was, among others, caused by the decline of epidemics resulting in mass death. Before the eighteenth century, 'the demographic impact of the profession of medicine remained negligible'(McNeill, 1976, p.239). It is presumed that ' deliberate inoculation of children with smallpox had been a folk practice in much of Asia for centuries, long before it came to the attention of European doctors and penetrated the repertory of their officially approved techniques in the course of the eighteenth century' (McNeill, p.253). It was introduced in England no earlier than 1721. Hygienic measures like the construction of sewers to prevent cholera and other epidemic diseases were only taken in the nineteenth century. This led to a decline of mortality among the inhabitants of cities in particular. McNeill describes this development as follows: 'By 1900, therefore, for the first time since cities had come into existence almost five thousand years previously, the world's urban populations became capable of maintaining themselves and even increasing numbers without depending on in-migration from the countryside.' (McNeill, p.275).

2.3.2 *Growth of energy consumption*

Economic growth since the industrial revolution was enabled by among other things the growing quantities of energy which became available for human use. Once food was the only source of energy for the only power which men had at their disposal: the power of their muscles. Other sources of energy became available for human use successively: fire (between 500,000 to one million years ago), the tractive power of animals, wood and charcoal, wind and hydroelectric power, fossil fuels and -for the last half century- nuclear fuel. World energy use increased sharply during the last century (figure 2.2.).

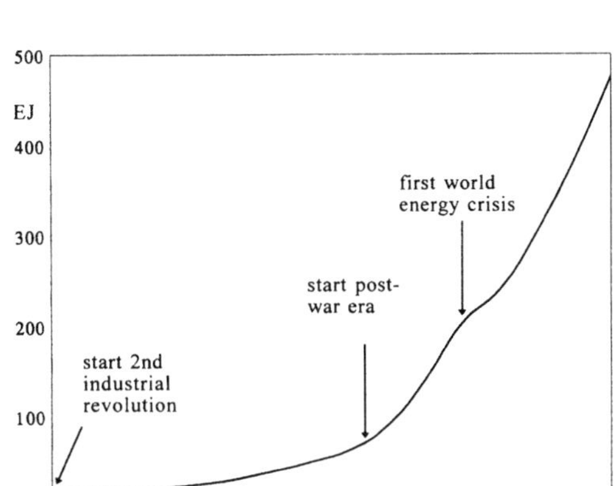

Figure 2.2 Growth of total world energy use (in EJ)
Source: Visser, 1987, p.1250 and OECD[b], 1993 p.22

2.3.3 Limits to growth

Time and again, modernization based on growing control and exploitation of nature, has provoked critical reactions. Since Rousseau, appeals for a return to nature were never completely absent in western society. Both radicals and conservatives were among the critics of modern society (See chapter 4 on Culture). Around 1970 there was a growing concern about the side effects of the growing use of natural resources.

In 1972 the Club of Rome, a group of prominent scientists, captains of industry and politicians, commissioned a report which was prepared by a group of researchers of the Massachusetts Institute of Technology headed by Dennis Meadows (Meadows *et al.*, 1972). According to the authors of this report, which attracted great attention world-wide, the limits to growth of human population and production would be reached within one hundred years. Scarcity of food, exhaustion of non-renewable resources and pollution were mentioned as factors limiting growth. The report's message was that it is better to limit the growth of population and production by purposeful measures than to have it halted by small or large catastrophes.

The report was strongly criticized both because of its questionable scientific quality and its political conclusions. It was concluded that there was no sufficient scientific evidence to justify definitive statements about absolute limits of growth of population and production. Scarcity of food is as much a distribution as a production problem and it cannot be foreseen to what degree technical progress will enable higher yields in agriculture. Neither is it possible to predict how many people will survive the increasing pollution. Stocks of available non-renewable resources will grow as a consequence of scarcity, increase in prices and technological innovation (See chapter 1). So for none of the three growth-limiting factors mentioned by the report, can any predictions be made as to if and when absolute limits of growth will be reached.

The report was also criticized because it completely denied the differences between rich and poor countries, between rich and poor people within countries and the conflicts of interests between them. However, it cannot be denied that the report 'Limits to growth', despite all its weaknesses, contributed substantially to the awareness of the enormous problems which are caused by the continuing growth of both population and production.

2.4 Growth and scarcity

In chapter 1 we discussed the complicated relationships between abiotic and biotic elements of the environment. Ecology is the scientific discipline which studies these relationships. Human induced change of one element of the environment will also have its impact on other elements. Commoner formulated 'the first law of ecology' as follows: 'Everything is connected to everything else' (Commoner, 1971, p.29). This 'law of ecology' conflicts with the striving for continued economic growth that is the political priority of modern industrial societies. From an ecological perspective, the strive for permanent growth of productivity of some elements of the environment implies the neglect of other elements of the environment. This insight is expressed by Commoner's fourth law of ecology as follows: 'There is no such thing as a free lunch' (Commoner, p.41).'In ecology, as in economics, the law is intended to warn that every gain is won at some cost' (Commoner, 1971, p.42).

In section 2 of this chapter we mentioned the restraint in land use by means of leaving arable land fallow in ancient times. Thanks to the availability of artificial fertilizers there is less necessity for restrain in land use nowadays. However, as a consequence of the use of artificial fertilizers, the concentration of nitrates in groundwater sometimes reaches a level that makes it unfit for consumption as drinking water. Modern industrial societies have developed great skill in partial control and exploitation of nature for specific human purposes. Time and again this partial use of nature leads to undesirable forms of exhaustion, pollution and disturbance of the environment. The partial control of nature is not only based on technical abilities but also on social power: on the power of the privileged to transfer unwanted 'external effects' of the use of the environment to the poor. We shall discuss this phenomenon more extensively in chapter 5 on social dilemmas and - as far as the spatial dimension is concerned - in the next chapter on geography.

Even when economic growth is not arrested by absolute, rigid boundaries, like those predicted in the report to the Club of Rome, it may result in 'new scarcity' (Hueting, 1980). Production processes often replace scarcity of consumer goods with scarcity of environmental quality. In so far as welfare also includes he availability of so-called 'free goods' like clean air and clean water, silence and the possibility for children to play in street, economic growth may also result in reduction of welfare. However, this is not expressed by the gross national product, which is considered the indicator of economic success par excellence.

Critics of the striving for continued economic growth are not new. As early as 1848 John Stuart Mill argued that growth of production of goods cannot go on indefinitely and that at the end a stationary state will be reached. He reproached his fellow economists for identifying economic growth as the one desirable goal. Since John Stuart Mill made his critical remarks about the continued striving for growth, bookshelves have been filled with critical comments on economic growth and its damaging effects on the environment. However, nowhere in the world any government has given up the striving for economic growth as a political priority for its own country.

A stationary state of wealth and population

If the earth must lose that great portion of its pleasantness which its owes to things that the unlimited increase of wealth and population would extirpate from it, for the mere purpose of enabling it to support a larger, but not a better or a happier population, I sincerely hope, for the sake of posterity, that they will be content to be stationary, long before necessity compels them to be.

It is scarcely necessary to remark that a stationary condition of capital and population implies no stationary state of human improvement. There would be as much scope as ever for all kinds of mental culture, and moral and social progress; as much room for improving the Art of Living, and much more likelihood of its being improved, when minds ceased to be engrossed by the art of getting on.

Mill, 1970, p.116.

2.5 Social limits

Much has been written about physical limits to the increase of welfare by means of economic growth. Less has been said about social limits to growth. According to Fred Hirsch (1977) there are physical limits to the reduction of scarcity if the quantity of coveted materials and products are limited. This is not only the case with stocks of non-renewable natural resources (see chapter 1, section 3.1.) but also with human products like the paintings of the Dutch old masters. Social limits to the reduction of scarcity by means of growth exist for both material products and non-material services. These social limits occur when the satisfaction offered by the individual disposition of an item is not based on its intrinsic properties but engendered by its scarcity. This is the case with goods and facilities which function as status symbols. 'Where the sole or main source of satisfaction derives from the symbol rather than the substance, this can be regarded as pure social scarcity' (Hirsch, pp.20-21). Even when satisfaction is derived from intrinsic properties of goods or facilities, there are social limits to growth. The side-effects of massive use may reduce satisfaction of individual consumers. Hirsch makes a distinction between 'physical' and 'social' congestion. A well-known example of physical congestion is offered by traffic-jams. If everybody drives a car part of the attraction of car driving disappears. An example of social congestion is offered by what may be called 'education inflation'. The school system functions as a selection process for a limited number of attractive positions on the labour market. Those with the most years of education are considered as the best qualified for the most attractive jobs. When more people are educated during longer periods individuals have to stay for more years at school to obtain positions which in the past were available after a shorter period of schooling. A certain level of education no longer offers the opportunities on the labour market that it offered in the past. Schematically, Hirsch summarizes the different forms of limits to the reduction of social scarcity as presented in figure 2.3.

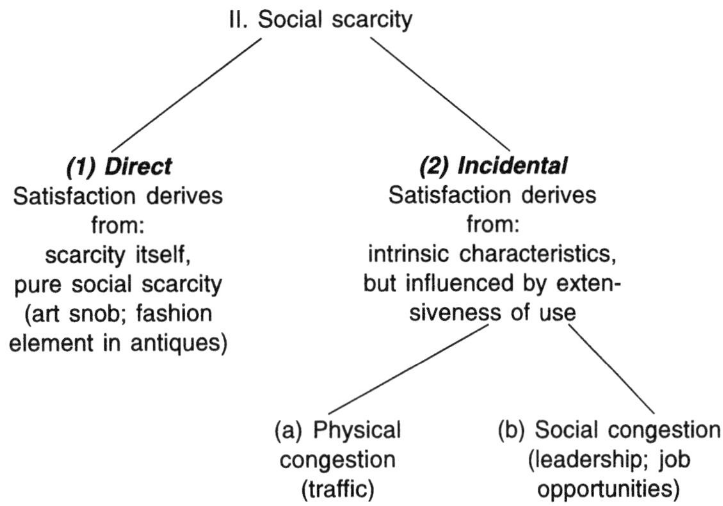

Figure 2.3 A categorization of consumption scarcity
Source: Hirsch (1977, p.21). Reprinted by permission. © Copyright 1977, Routledge and Kegan Paul.

Hirsch distinguishes between the 'material economy' and the 'positional economy'. The *material economy* includes goods and services of which the intrinsic value or usefulness does not suffer from growth of production, although 'external effects' may reduce the satisfaction of other needs of 'free goods' like fresh air and silence. The striving for more welfare by means of economic growth is based on the presuppositions of the material economy. According to Hirsch, there also exists a *positional economy* consisting of goods and services whose value for individual consumers declines in the case of growing delivery, because satisfaction is dependent on its scarcity.

Hirsch presents the distinction between material and positional economy as a distinction between different goods and services. In reality however, this is a distinction between different meanings of goods and services for owners and consumers. Every single item can have both a 'material' and a 'positional' meaning for its owner or consumer.

According to Hirsch, in societies in which the most vital needs, like the need for food, clothing and housing, are satisfied, the positional aspect of goods will gain in prominence. He wonders 'what happens when the material pie grows while the positional economy remains confined to a fixed size?' (Hirsch, p.27).

In our own words: what happens when more foreign travel is experienced and more years are spent in school and universities, but the number of people who can excel by recounting an extraordinary travellers tale or who can obtain an attractive job remains limited? Answer: these happy few will have to travel further and further and to study for longer and longer periods. Again in the words of Hirsch: 'For the individual, increased positional competition involves a deterioration in the social environment. More individual effort and resources have to be expended to achieve the same result' (Hirsch, 1977, p.53). The competition in the positional economy leads to unlimited needs for goods from the material economy.

The attraction of Hirsch's analysis of the phenomenon of scarcity consists of the insight it offers into both the causes of environmentally damaging growth of consumption and the expansion of educational facilities and other social services which are no longer affordable.

We have discussed in general terms modern industrial society and how it originated in England and then spread across the world. In the next chapter we will discuss differences in development between different parts of the world.

2.6 The information revolution

Although contemporary western society is still often referred to as an 'industrial' or 'industrial-capitalistic' society, the place of industry is becoming less and less dominant. As in the past employment moved from the agricultural to the industrial sector, in recent decades employment has shifted from the industrial sector to the sectors of commercial and non-commercial services.

Only two centuries ago, according to our school-books, the industrial revolution took place. Nevertheless it is highly probable that future historians will describe the second half of the twentieth century as the era in which, in industrial societies, the information revolution took place, or rather the transition from an industrial to an informational society. Again we are experiencing a change in means of production. In comparison with the processing and distribution of materials, the making and exchange of information is gaining in prominence. As a symbol of economic activity the factory building is replaced by the open-plan office with viewing screens, or even by the living room of the tele-worker, with the same equipment.

This development has a great influence on the environmental impact of enterprises. In general, the making and the spreading of information will lead to less environmental stress than the production and distribution of material goods. What is more, information technology contributes to the control and steering of processes which might pollute the environment. In the next chapter we will examine this development more extensively in discussing the concept of 'ecological modernization'.

Chapter 3
Geography

3.1 Variety and scale

In Mexico City the local population suffers from extreme air pollution. A large part of the bottom of the Aral Sea in central Asia stands clear of water and fishermen are no longer able to do their work. In Malaysia the indigenous population struggles for the conservation of its habitat in the tropical rainforest. In the Netherlands people near Schiphol airport suffer from the noise of aeroplanes which disturbs their night's rest.

Environmental problems exist all over the world. In the preceding chapter we dealt with environmental problems in a historical perspective. In this chapter we will discuss the geographical variety of environmental problems. Time and space are important dimensions of environmental problems in another way. Damaging effects of human-induced environmental change can manifest themselves after shorter or longer periods and at shorter and longer distances. Different spatial and time scales in which damaging environmental effects of human activities become manifest and remain so will be the subject of the second part of this chapter.

3.2 Geographical variety

Not long ago a map of the world was published on which different colours indicated environmentally undisturbed, partially disturbed and disturbed regions of the world. According to this map Europe, North America and South-east Asia are the most disturbed parts of the world (Hannah, *et al.*, 1994). In reality the variety of forms of human-induced environmental decay in different parts of the world is enormous. This variety is partly caused by differing natural circumstances. Soils in tropical countries are in general more sensitive to erosion than soils in Europe and North America. Air and water pollution have much stronger local effects in extremely cold Siberia than in the temperate zones of the earth. Not only on a world-wide scale but even on a much smaller scale differences in natural circumstances strongly influence the effects of pollution. For instance, in a small country like the Netherlands the effect of acid deposition is much stronger on the sandy soils in the central and southern parts of the country than on the clay soils in the western and northern provinces.

Geographical variety in environmental disturbance is also influenced by differences in the three main factors contributing to environmental problems which were mentioned in the preceding chapter: population size and quantity and quality of environmental utilisation.
In this chapter we will return to these three main causes of environmental problems. But we start from another perspective: the well-known division of the world into a first, second and third world. Although these terms are becoming outdated, the division is still a good base from which to approach the geographical variety of environmental problems in different parts of the world. In this chapter our aim is to offer some insight into the environmental effects of basic differences between socio-economic systems. The difference between developed market economies (first world), (former) planned economies transforming into market economies (second world) and developing countries with marginally developed market and state regulation (third world) will still be crucial for the understanding of both the existence of environmental problems and the possibilities of environmental problem-solving in different parts of the world over many years.

3.2.1 Market economies

Our characterization of market economies refers in particular to the comparatively rich countries which are united in the Organisation of Economic Cooperation and Development (OECD). To this organisation belong many, mainly western, European countries, the USA, Canada, Mexico, Australia, New Zealand, Japan and South-Korea. Recently the Czech Republic, Hungary and Poland have become members of OECD.
In market economies demand and supply of goods and services are dependent of prices. According to the logic of the market-mechanism, growing scarcity of environmental goods would have to lead to an increase in prices and reduction of scarcity of these goods. In reality however, scarcity of environmental goods is often not expressed in prices (see section 5.2). Several scarce environmental goods like clean air and water, silence and scenic beauty are not at all protected by the market mechanism because they are not priced. In section 1.3.1. the exhaustion of non-renewable and renewable resources was discussed. Predictions about the moment of exhaustion of non-renewable resources are risky. However, it is likely that the massive use of all sorts of materials and fuels will lead to exhaustion or strongly reduced availability of these environmental goods within one or a few centuries. Nevertheless, this scarcity is not expressed by the current prices. Populations of plants and animals can be reduced or disappear as a result of human behaviour both in a direct and an indirect way. In a direct way by excessive harvesting or hunting, in an indirect way because the living conditions of animals and plants are destroyed. In this case the market mechanism does not prevent environmental destruction either.

This characterization of market economies would be incomplete without mentioning the fact that the market economies of the OECD countries have developed in the context of a political system which is able to correct negative environmental effects of individual behaviour and institutionalized activities. In Europe, processes of state formation started in the sixteenth century, long before the industrial revolution. In the nineteenth century control of industrial pollution became a major state concern. Since that time authorities at national, regional and local level have contributed to the reduction of pollution and stimulated a healthy environment by means of sewers, water purification plants and garbage collection. In chapter 9 we will discuss extensively environmental tasks of states. Here we will only make a few remarks about the possibilities and limitations of market economies in solving environmental problems. In the rich countries with market economies, the late sixties and the early seventies were a period of growing concern about decay of environmental quality both locally and globally. Since that period activities of individual and organized citizens, public authorities and private enterprises have been successful in reducing some forms of environmental stress. In particular air and surface water have, as far as several pollutants are concerned, become cleaner. In England, thanks to a drastic reduction of sulphur dioxide emissions, the type of air pollution ('smog') that still killed thousands of people in 1952 no longer occurs. The salmon has returned to the river Thames.

In recent decades progress has been made in the reduction of both use of scarce natural resources and emission of polluting substances. In general industrial production processes have become more efficient in the use of materials and energy. Not only production processes, but also products have become more material – and energy-efficient.

As a result of the developments mentioned above modern industrial production in market-economies seems to become cleaner and more efficient in its use of natural resources. In the eighties and nineties, environmentally induced criticism of the basic institutions of western societies and their continuous drive for economic growth are gradually replaced by a growing trust in the possibilities of the institutions of market economies to solve environmental problems. The concept of 'ecological modernization' (Huber, 1982; Mol, 1995) has become popular as an expression of this optimism.

However, there are reasons to tone down this optimism. Recently it has been suggested that the so-called process of 'delinking' of environmental pressure from economic growth has already been succeeded by a process of 'relinking' in some advanced economies (De Bruyn en Opschoor, 1997). Successes have been achieved in the reduction of pollution of air, water and soil caused by industrial production processes. But the continuous growth of consumption of industrial products has hardly been influenced by environmental protection measures. Waste streams of used products and their remnants are still growing. The exhaustion of natural resources and the disappearance of species of animals and plants goes on.

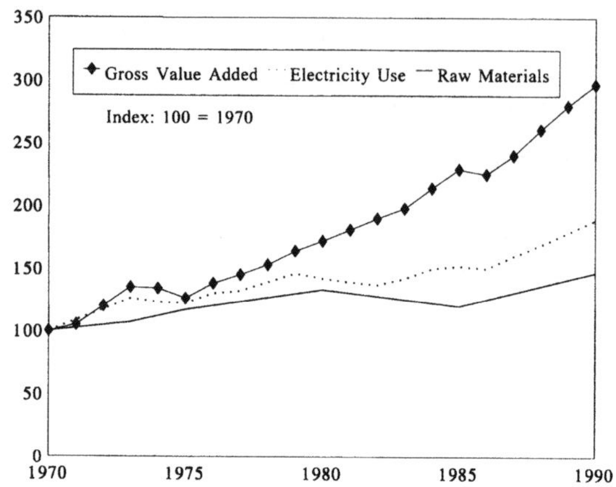

Figure 3.1 Use of some selected resources in Japan compared to economic growth Source: (Jänicke *et al.*, 1993).

Progress has been made in the reduction of directly visible pollution of the everyday surroundings of human beings. The difference in reduction of surface water and ground water pollution is typical of the relevance of visibility of pollution. Pollution of the visible surface water in rivers and lakes in general has been reduced but pollution of invisible ground water goes on. In section 1.3.2. we distinguished between flux-type and sink-type pollution. In countries with a market economy the latter type of pollution goes on, at best more slowly. Nowadays environmental problems in market economies are in particular symptoms of affluence. Progress as a consequence of cleaner and more efficient production processes and products is often partially or completely counteracted by the effects of growing quantities of consumer products and growing use of these products. An example of this trend is offered by the transport sector (See text below in frame).

Energy use in transport

In a number of OECD countries (Japan, USA, (western) Germany, France and the United Kingdom) total energy use of four types of transport (cars, buses, trains and airplanes) has grown in the period 1970 - 1987. This growth was not only caused by increasing mobility leading to a growing number of passengers/kilometres. The increase in the number of cars goes hand in hand with a decrease of the number of passengers per drive. As a consequence growing efficiency in energy use per car/kilometre can be coupled with growing energy use per passenger/kilometre. In all researched countries (with the exception of the United States) energy use per passenger/kilometre indeed increased. The positive effects of technical improvements in efficiency were completely nullified by the reduced number of passengers per drive. On top of that there is the effect of the growing number of passenger/kilometres.

Schipper and Meyers, 1992.

Unlimited growth of production and consumption
Market economies in general have gone through a process of demographic transition. Old stability or slow growth of population, based on both high birth and death rates, came to an end when, thanks to the progress of health care and environmental hygiene, numbers of death declined. This change was followed by the decline of birth rates. The damaging environmental effects of production processes and products have been reduced in many cases.
In chapter 6 we will discuss the tendency for consumers who gain money by restraining from certain polluting types of consumption to spend the money saved on other equally or even more environmentally damaging forms of consumption. An example of that phenomenon is offered by persons who dispose of their private cars and instead of using a private car make use of aeroplanes to reach their holiday destinations. Growth of production and consumption partly or (more than) completely nullify the effects of more efficient and cleaner production processes and products. It is noticeable that in case of the renewable resources of fish, quotas have been assigned but that the use of the non-renewable resources of industrial production is not restricted in any way. In countries with a market-economy environmental policy is unable to counteract the long-term and world-wide damaging effects of growing production and consumption.

3.2.2 Planned economies

In the twentieth century large parts of the world were governed by communist regimes for shorter or longer periods. This was, among others, the case in the fifteen republics which once belonged to the Soviet Union and in other countries in central and eastern Europe, in China, Cuba, North Korea and Vietnam. These countries are now in a process of more or less radical transformation into a market economy. Yet the environmental impact of former and still existing communist regimes is still impressive and will remain so in the coming decades. Beside that, for all those who are interested in the environmental effects of political systems the planned economy of (former) communist countries remains until now the only alternative to the market economy which has been put into practice on a large scale in industrial countries.
In section 2.2.4. Marx, the ideological founding father of communist planned economies, was quoted while praising the achievements of capitalism. For Marx capitalism was a necessary phase between feudal and communist society. Capitalism started the development of industrial technology and management but communism would surpass the industrial achievements of capitalism. In future classless societies not only the exploitation of the working class by capitalists, but also the so-called 'anarchy of production', with its alternate periods of over-production and stagnation, would come to an end. Production for profit would be replaced by production for the fulfilment of basic needs.
In planned economies production targets are not regulated by the market-

mechanism but by means of central planning. State bodies instead of innumerable entrepreneurs decide what will be produced and how. In theory this enables rational decision-making about production targets and limitation of damaging environmental effects of production processes and products. In reality planned economies were unable to fulfil basic needs of citizens. What is more, this failure to fulfil the needs of consumers was accompanied by extreme forms of environmental pollution and destruction of nature. Notorious examples of environmental catastrophes in this part of the world are the shrinking of the Aral Sea (Wolfson, 1994), the explosion in reactor 4 of the nuclear power plant of Chernobyl on April 26, 1986 and other nuclear accidents in the former Soviet Union (Medvedev, 1990), and the decline of cultivated land as a consequence of erosion and desertification in densely populated China (Smil, 1994, p.57).

Nuclear nightmares

Chernobyl is only the most dramatic manifestation of the post-Soviet nuclear military-industrial complex. Other manifestations are:

- The long-lasting effects of nuclear explosions for non-military purposes
- Radioactive waste from radiochemical industries
- Accidents with atomic submarines
- Radiation from radioactive waste, reactors and other materials, dumped in the sea
- The risks of the use of plutonium by terrorists
- New accidents with nuclear power stations.

Causes of environmental problems in (former) communist states

In the next chapter on culture we will discuss Marxist ideology as a source of environmental destruction in communist countries. Other causes of environmental problems in (former) communist countries have to do with the (former) economic and political system of these countries.

The economy

- The modest role of prices.

Like market economies, planned economies tend to underestimate the value of natural resources. Planned economies are even more extreme in denying the value of nature than market economies. According to Marx's theory of labour value, natural resources only have use value, but no exchange value, as expressed in prices. Exchange value derives from human labour invested in production. The free, unprized availability of natural resources has often been considered, both by supporters and opponents of the communist states, an essential feature of this type of society. Nowadays the free or cheap availability of resources is often mentioned as a cause of wastage. Examples are the tendency constantly to relocate mines in order to obtain easily accessible minerals which is not tempered by financial barriers; the excessive irrigation of agricultural grounds by distribution of water free of charge leading among other things to the above mentioned shrinking of the Aral Sea and salination of agricultural grounds and the large quantities of materials and energy used in industrial production. In the delivery of raw materials and energy to enterprises prices often hardly played any role. Good contacts with officials who decided on the delivery of these resources to enterprises were often of more importance than money.

– The irrationality of planned targets.

In centrally planned economies planned targets were often formulated in physical terms, like the quantity of used material inputs. Therefore entrepreneurs were stimulated to use large quantities of materials. For many years absurd results of the physical planning indicators have not only been described in western publications but also in the Soviet press (Tellegen, 1989). As a result of the use of physical planning indicators sometimes only big pans or thick pieces of paper were available because by producing small pans and thin pieces of paper it would take more time and effort to fulfil planned targets.

The political system

Environmental problems are mostly 'external' effects of activities of widely different types of organisations. In general these activities can not be 'abolished', but have to be moderated or changed in order to avoid environmental damage. Industrial production, motorized transport or housing cannot be 'abolished', but they can be limited, at least in certain areas, or changed. Outward 'countervailing power' is necessary in general to change the behaviour of organisations to avoid or limit environmental damage. In the former communist states 'countervailing power' was weakly developed. The secrecy about the Chernobyl accident and other nuclear accidents is a typical example of a political strategy of avoidance of 'countervailing power'. There existed an institutional division between 'government' and 'party', but this did not generate 'countervailing power' to avoid damaging environmental effects of governmental decisions. Environmental control was integrated within the ministries for different branches of industries whose main task was the fulfilment of planned targets. There was no free press and no freedom to form action groups and environmental organizations. Therefore, the scope for environmental protest of individual or organized citizens was small. Environmental protest was also hindered by the strong local power of industrial enterprises. They did not only provide jobs but also housing and even the construction and exploitation of cultural and recreational facilities. Local populations were dependent on local enterprises for much more than just their employment.

No ecological modernization

Populations in central and eastern Europe including the former Soviet Union are growing slowly, remaining stable or, as in Russia, even shrinking. In densely populated China tremendous efforts have been made to reduce population growth. In the early seventies the wan xi shao ('later, longer, fewer') campaign aimed at postponement of marriage, lengthening of intervals between births and limiting family size to two children (Smil, 1994, p.19). At the end of the seventies the more radical one-child campaign started.

Chinese population policy

The one-child campaign started in January 1979, and its effectiveness rested on a combination of rewards and penalties. Rewards for one-child families included cash bonuses, preferential housing, job, and school enrollments, and free medical care and schooling for the child in cities. For peasants the rewards were less appealing consisting mainly of additional communal workpoints and larger private plot and grain allocations. Penalties for more than two children included withdrawal of health subsidies and paid maternity leaves, reduced rations for nonstaple foodstuffs, and workpoint reductions'.

Smil, 1994, p.22 Reprinted by permission.

In China the demographic transition from high to low birth and death rates took place in the thirty years between the forties and seventies of the twentieth century. This is a very short period compared with the period of approximately hundred years in which this process took place in several European countries. Yet until now population policy in China has not been successful in reaching its targets for maximum future population size. At the end of the nineties there will be 125 million more Chinese than at the beginning of this decade (Smil, 1994, pp.19, 20 and p.34).

In the transforming former communist countries of central and eastern Europe decline of production has contributed to decline of some forms of environmental pollution in recent years. The quantity of consumer products is, to put it mildly, quite moderate, compared with market economies. The basic environmental weakness of (former) communist states lies in the quality of production, including mining and transport. Production processes are outmoded and squandering and environmental control on wasteful and polluting production processes is weak. In discussing market economies we used the concept of 'ecological modernization' and referred to environmentally relevant improvements of industrial production in countries with market economies in the past decades. These types of improvements did not take place to that degree in the (former) communist countries in the same period. In general planned economies were less efficient in the use of raw materials and energy than market economies. Lack of efficiency of the Soviet economy was one of the reasons for starting with *glasnost* and *perestroika* in the former Soviet Union (Tellegen, 1989). Since the fall of the communist system environmental investments and environmental control have even declined in some of these countries. In (former) communist states there are still relatively large possibilities for reduction of wastage and pollution in the production processes and transport of goods produced to the consumers. In theory the transformation towards a market economy and growth of production and consumption can be combined with such improvements in efficiency.

3.2.3 Developing countries

Many societies in the world, in particular in the southern hemisphere, cannot be called market economies or planned economies. They are called 'developing countries' although 'stagnation' and 'decay' are better designations of the situation in some of the southern states than 'development'. Developing countries can be found in Africa, Central and Southern America, Asia and Oceania. Our characterization of these countries refers in particular to Sub-Saharan Africa. The economically inspired division between market-, planned- and developing economies parallels a technologically inspired division between post-industrial, industrial and pre-industrial countries. In the rich market economies the information sector is growing and the industrial sector is shrinking. In the (former) planned economies this stage of development has not yet been reached. The poor countries of the southern hemisphere have not yet gone through a process of industrialization. Larger parts of the population can only survive by means of gathering food and firewood and subsistence farming.

The vicious circle of population growth, environmental destruction and poverty in Sub-Saharan Africa

Rapid population growth, environmental degradation, and slow agricultural growth in Sub-Saharan Africa are closely linked. The principal problem is that the technologies applied in shifting cultivation and transhumance pastoralist systems, appropriate under conditions of low population density on Africa's fragile natural resource base, are environmentally damaging when practiced by rapidly increasing populations. When population densities increase and shifting around on the land becomes impossible, but farming practices do not change, soils degrade and forests are destroyed. Soil degradation and deforestation constrain agricultural growth. Lagging agricultural growth perpetuates rural poverty and food insecurity, which in turn impede the onset of the demographic transition to lower human fertility rates.

Cleaver and Schreiber, 1994, p.198.

In section 2.3 we discussed the fall of pre-industrial civilizations as a result of internal dynamics, in particular soil degradation. Environmental degradation in contemporary developing countries often can be ascribed to 'fragmentary modernization' (Mansvelt Beck, 1981, p.82), which means: the penetration of western technologies, forms of organisation and cultural values in pre-industrial societies. This influence can lead to population growth (as a result of the introduction of western health care), construction of roads, dams and storage reservoirs, introduction of new types of agriculture (including monocultures) and destruction of tropical forests. Whereas in market economies and planned economies modernization developed as an internal process, in developing countries technological and cultural changes were 'imported' from abroad both during and after the colonial era.

Typical of fragmented modernization in developing countries is the side by side existence of a market economy and a traditional economy. Often there are resource-use conflicts between the competing demands of cash crops and food crops' (Moyo *et al.*, 1993, p.274). The market economy is oriented to export to the world market. Export often leads to destruction of nature, in particular tropical forests, and selling out of raw materials at low world market prices. But the subsistence economy, not regulated by market or state, also contributes to environmental disturbance. Environmental problems in developing countries are both a consequence of production for export and for the local population.

The Hamburger connection: myth and present reality

The development of cattle rearing has been a major cause of loss of tropical forests. In Costa Rica in the pre-1980 period forest-to-pasture conversion led to more deforestation than all other economic activities, including commercial logging, together. For that reason Costa Rica was considered as an extreme example of the so-called 'hamburger-connection' or: deforestation in poor countries to feed northern Americans and other citizens of rich countries with hamburgers. But in the 1980s circumstances drastically changed. 'At present however, the rainforest-North American hamburger connection is questionable in a literal sense for 2 reasons:

- overall beef exports from Costa Rica declined in the 1980s as compared to the 1970s
- domestic consumption of beef in Costa Rica rose during the same period of comparison (....)

By 1989, 76 percent of all beef produced was consumed internally, while the remaining 24 percent of total production was exported to foreign markets'

Lehmann, 1992, pp.62-63.

Compared with other parts of the world the number of inhabitants per square kilometre in developing countries in general and Sub-Saharan Africa in particular cannot be called high. It is, among other things, the combination of underdeveloped agricultural technology and organisation and unequal division and accessibility of agricultural grounds that leads to both excessive exploitation of natural resources and food scarcity. Only in combination with these factors can population growth and density be called a major cause of environmental problems in Africa. Resource scarcity in Sub-Saharan Africa is a cause of mass starvation, mass immigration and violent environmental conflicts. Here and elsewhere environmental problems are becoming sources of political tensions, military conflicts and sometimes even genocide.

Population density and genocide in Rwanda

Between 1970 and 1989 the population of Rwanda grew from 3.756.000 to 7.128.000 and the population density from 143 to 270 per square kilometre.

'Population density can attain incredible heights in certain localities such as Shyanda *commune* where it had reached 668 people per square km. by 1989. Population projections for the year 2000 topped 10.000.000 people, with a 50% increase every ten years leading to over 50.000.000 by 2040 i.e. twenty times the population density of France, unless there is some geographic slowing-down in the mean time. Grim as it may seem, the genocidal violence of the spring of 1994 can be partly attributed to that population density. The decision to kill was of course made by politicians, for political reasons. But at least part of the reason why it was carried out so thoroughly by the ordinary rank-and-file peasants in their *ingo* was the feeling that there were too many people on too little land, and that with a reduction in their numbers, there would be more for the survivors'

Prunier, 1995, p.4.

No demographic transition

A major difference between market and planned economies on the one hand and developing countries on the other hand is that the former have more or less completely gone through the process of demographic transition, whereas this process has not yet started or is only in the beginning in the developing countries. However, there is a great variety among what are called here 'developing' countries and this is in particular the case with the development of population sizes. Several developing countries (among them India and Indonesia) were more or less successful in their strive for declining population growth. On-going population growth in these countries can be explained by the fact that the decline in birth rates stays behind the decline in death rates.

Problems of prosperity and poverty

The developing countries differ from the rich market economies, because their environmental problems are problems of scarcity instead of problems of affluence. Whereas in rich countries the still growing quantities of consumption are a major cause of environmental problems, in poor countries it is the scarcity of food and firewood that contributes to excessive exploitation of natural resources.

3.2.4 Societies in transition

Paraphrasing George Orwell, it can be said that all societies are in transition, but some are more in transition than others. The latter can be said of former communist and some of the developing countries.

In former communist states the closure of factories and the drastic reduction of industrial output has led to a reduction of environmental pollution. However, if a long-term reduction in environmental damage caused by productive activities is to be achieved, existing facilities will need to be modernised and particular economic activities will have to be replaced by others (Tellegen, 1996, p.87). Environmental control was a weak element in former communist states, but after the fall of communism previously existing forms of environmental control disappeared without their being replaced by new, effective forms of environmental control (Tellegen, 1996, p.88). The introduction of the market economy also leads to a growing use of private cars and a deterioration of public transport systems. Recyclables are replaced by throw-away packaging. It is to be expected that the development of environmental policy measures will for many years lag behind the development of new types of production and consumption in these countries.

In particular in south east Asia the so-called 'new industrial states' (Harris, 1990) are successful in their strive for high economic growth percentages. This success is accompanied by excessive growth of pollution.

Smog over East Asia

Boomtime East Asia has gotten a whiff of the future - and it is acrid and dark. A pall of smog has decended over Singapore and much of Malaysia in recent weeks. It is the latest sign of a spreading epidemic of pollution that is eroding the quality of life in many parts of East Asia even faster than incomes and material living standards rise, scientists and researchers say. Indeed, even Singapore, which prides itself on having a clean and green environment, is becoming a victim of East Asia's supercharged economic growth (...).Pollution in Singapore and Malaysia from motor vehicles, industry and power stations normally disperse quickly. But in recent weeks it has been kept in place by layers of smoke from forest fires that started in August in neighbouring Indonesia. The fires are caused by excessive commercial logging, slash-and-burn agriculture by poor farmers, conflicts pitting labourers and local land users against timber and plantation companies, and tinder-dry conditions. Until the smoke blanket recedes, Singaporeans are having to adjust their lifestyles. The government issues pollution updates three times a day. Hospitals and medical clinics have reported a rise in respiratory complaints and skin and eye irritation cases.

International Herald Tribune, September 30, 1994.

As is clearly illustrated by the case of air pollution described in Singapore environmental problems are more and more becoming transboundary and global phenomena. They can less and less be solved in only one country or even one part of the world. There is a growing interdependency between the different types of societies which we discussed in this section.

3.3 Spatial scale of environmental problems

So far we compared environmental problems in different parts of the world. Another geographical dimension of environmental problems is the spatial scale on which they are caused and experienced. The noise of the bar by which the neighbours are annoyed cannot be heard further away than a few blocks at most. The use of CFK's containing sprays contributes, by damaging the ozone layer, to environmental disturbances all over the world. Human beings can contribute to and suffer from small scale and large scale environmental changes. The 'global' or 'world' scale has often been considered as the largest scale level of human-induced environmental disturbances. But pollution of the extra-terrestrial space is already becoming an urgent problem. The concept of debris in space has recently entered the environmental vocabulary. At the end of the twentieth century environmental pollution caused by human activities exceeds the global scale level.

Five spatial levels of scale on which environmental problems become manifest have been distinguished: the local (or 'inner region') scale, the regional (or 'lake/rural') scale, the fluvial (or 'sea/fluvial') scale, the continental (or 'ocean/continent') scale and the global (or 'planet') scale. For each of these levels some characteristic processes and problems have been mentioned (Langeweg *et al.*, 1989). They are presented in figure 3.2.

In figure 3.2 pollution of soil, water and air are related to ascending scale levels. Indeed, the scale level at which problematic consequences of human behaviour become manifest is related to the state of aggregation of substances.

In general *soil and solid substances* spread less than liquid substances and liquid substances do so less than gaseous substances. In general, the ease with which substances spread is inversely proportional to the visible effect of their spread. Solid substances which are removed do not change in size and composition. Removal can be considered as beneficial both at the place of departure and arrival. In the past citizens as well as farmers profited from the dumping of urban household waste on agricultural soils. Because of its composition, household waste contributed to soil improvement. When after the introduction of sewage systems faeces were separated from the remaining household waste and when in the remaining waste the share of non-degradable waste substances was augmented, urban household waste lost its significance as a contributor to soil improvement. Moreover the availability of artificial fertilizers reduced the need for compost made of urban waste. A comparable development has taken place with harbour sludge. In the past sludge that was released by dredging harbours

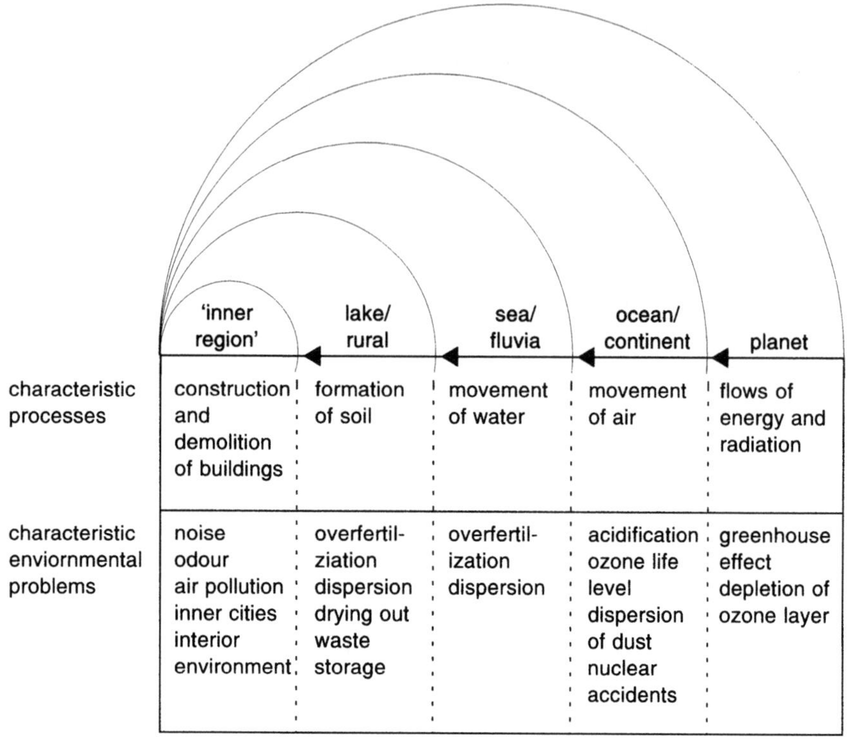

Figure 3.2 Spatial scales of environmental problems; characteristics and problems at the global, continental, fluvial, regional and local level
Source: Langeweg *et al.* (1989, p.4.)

and rivers could be used to clear land for building. Nowadays this sludge is so heavily polluted by chemical substances that it has to be dumped as waste.

As for the state of aggregation *air* is the opposite of soil and solid substances. There has always been a great contrast in public regulation of soil and air. In the course of history more and more parts of the earth's surface became privately owned. Only in communist countries developments went in the direction of shrinking and disappearing private ownership of land. On the other hand air was freely available for everybody and free of regulation.

Of course, reality is somewhat more complex than the statements above suggest. At the local level, air pollution has been experienced as a nuisance for a long time. But at higher scale levels (see figure 3.2) air was the 'free good' par excellence. This changed when it was discovered that emissions of polluted air in the United Kingdom and other western European countries contributed to the acidification of Scandinavian lakes. At the UN environmental conference in Stockholm in 1972 the phenomenon of transboundary air pollution was put on the international environmental policy agenda. Since 1985 western-and eastern European countries have made agreements to reduce emissions of pollutants

contributing to acidification (see section 10.1).
Simply put, in the past pollution by solid and gaseous substances only caused local environmental problems. In the case of solid substances because of the immobility of these substances, in the case of gaseous substances because of dilution as a result of their great mobility.
As regards governmental control water differs from both soil and air. Because of its mobility and in contrast to land, water often cannot be exclusively possessed by single public or private owners (Kinnersley, 1988). Water flows through rivers and seas and moves by evaporation and condensation in cycles through air (clouds) and soil (vegetation, ground water). On the other hand even in the past water was not a 'free good' like air. In ancient times the use of water had already led to common efforts and arrangements. In section 2.2 irrigation was mentioned as an example of such common efforts. In the Netherlands water management was the first subject of formalized democratic decision-making. Common arrangements were and are necessary to reconcile different and often conflicting interests. Fisheries, shipping trade, the use of water-power as energy source, land reclamation and protection against floods, irrigation, drinking water supply and discharge of waste all require specific conditions and all have specific effects on water-quantity and/or quality which can come in conflict with each other (Kinnersley, 1988, pp.24-33). For that reason regulation of water has a long history.
Whereas land is either private or public property and whereas air is free available, water is the domain of common efforts and regulations for the benefit of different users and their interests.

3.4 Space and time

Between the beginning and the solution of environmental problems, three phases have been distinguished:
- the *initial period*, from the time that an intervention (in a flow) is greater than the natural interference, until the time that undesirable effects manifest themselves;
- the *regulation period*, from recognition of the problem until implementation of the measures;
- the *recovery period*, to be subdivided into a natural recovery time and the time required for restoration' (Langeweg *et al.*, 1989, p.6).

The natural recovery time of a local problem like noise pollution is no longer than the time in which the sound moves from its source to its receivers. When the radio is turned off or the machine is stopped the noise pollution is over practically immediately. With a global problem like the enhanced greenhouse effect this is quite different. Even if all emissions of greenhouse gases were to come to an end today (which is absolutely impossible) the greenhouse gases emitted earlier are still there. The initial and the regulation period of the enhanced greenhouse effect are also longer than those of noise pollution. Sound effects manifest themselves immediately, although it may take a long time before they are experienced as annoying. Increase of concentrations of greenhouse gases as a result of human activities is only noticeable after a long period. Even after the increase has been recognized it still takes a long time to reduce the emissions of the many different sources in different countries which contribute to the greenhouse effect. Because of the impossibility of influencing many different human activities all over the world, it is not very likely that internationally agreed targets of reduction of greenhouse emissions will be reached. In chapter 10 we will come back to this issue.

The correlation between spatial and time scales of environmental problems is far from perfect. A local environmental problem like soil pollution can have an initial period of several centuries. This is, for example, the case with lead pollution caused by potters many centuries ago. In this case the long initial period cannot be attributed to slowness of the pollution process but has to be explained by the growing sensitivity to this type of pollution in recent years. Not every discovery of soil pollution will lead to immediate removal. Therefore the regulation time may be long. For some cases of soil pollution which are already known, removal will even not be started, let alone finished, within a few centuries.

The regulation time of local environmental problems like noise pollution is dependent on the power of those who cause the noise. A local barkeeper will in general be less powerful to resist environmental protection measures than an international airport. In the latter case the continuation of noise during sleeping periods of surrounding citizens can be justified by the necessity to maintain and, if possible, to increase the share in international transport and the interests of the workers employed by the airport.

An example of a small scale environmental problem with a long recovery period is offered by groundwater pollution caused, among other things, by artificial fertilizers and pesticides used in agriculture. In this case the slowness is not primarily caused by social factors, like late discovery and development of awareness of the problem, but by physical circumstances. Both the initial and the recovery period can be very long. After polluting emissions have come to an end, it may take still hundreds of years before the groundwater becomes clean again.

We may conclude that the time-scale of environmental problems not only depends on their physical properties but also on their social context. We will deal with this matter more extensively in chapter 7. Spatial and temporal scale of environmental problems are relatively independent of each other. Locally caused and locally felt environmental problems with a short initial period generally will be solved sooner and more effectively than problems with a long initial period and with global impacts. However, there are also environmental problems with quite different spatial and time scales. In figure 3.3 this is illustrated with some examples.

Time scale \ Spatial Scale	***Small***	***Large***
Short	Noise	Contamination vegetables by the Chernobyl nuclear accident
Long	Groundwater pollution	Enhanced greenhouse effect

Figure 3.3 Time scale and spatial scale of some environmental problems.

3.5 Direct and indirect transboundary effects

Transfer to others of harmful side effects of the use of environmental resources has been considered as the core of environmental problems. We will discuss this issue extensively in chapter 5. This phenomenon also has a spatial dimension. Damaging effects of environmental use in one place may spread by air and water to the farthest corners of the earth, including Antarctica and the bottom of oceans. From a political viewpoint there is a great difference between transboundary and global environmental effects. In the latter case different parts of the world contribute to common, global problems. On a world-wide scale measures can be taken to solve or at least to mitigate global problems. We will discuss global problems and global environmental policies in chapter 10. In case of transboundary effects, environmental problems in one country or a group of countries are hurting the interests of one or more other countries. Two types of transboundary effects of environmental damage can be distinguished: direct effects and indirect socio-economic effects.

A. Direct socio-economic effects

Emissions of polluted material in air and surface water and transport of hazardous waste can have damaging effects on nature and people in other countries. The same can be said of accidents with nuclear power plants or oil tankers. Not only local environmental damage, but also local use of scarce resources can have transboundary repercussions. In particular scarcity of river water can lead to conflicts between states.

Scarcity of river water and military conflicts

'Our research suggests that the renewable resource most likely to stimulate interstate resource war is river water. Water is a critical resource for personal and national survival; furthermore, since river water flows from one area to another, one country's access can be affected by another's actions. Conflict is most probable when a downstream riparian is highly dependent on river water and is strong in comparison to upstream riparians. Downstream riparians often fear that their upstream neighbours will use water as a means of coercion. This situation is particularly dangerous if the downstream country also believes it has the military power to rectify the situation. The relationships between South Africa and Lesotho and between Egypt and Ethiopia have this character'

Homer-Dixon, 1994, pp.5-40

Water scarcity always has played a major role in the relationships between Israel and its neighbour states. In this case not only use of drinking water but also the transboundary effects of use of groundwater are a source of political tensions. 'Water diplomacy' is a means to lessen tensions and to avoid conflict (Dolatyar, 1995).

B. Indirect socio-economic effects

Local environmental deterioration can have indirect socio-economic effects in other areas. One of these effects is mass migration of people from environmentally damaged places to other countries. It is not always clear to what degree environmental disturbance actually is the only or main motive for migration. Illustrative of this uncertainty is the immigration from Belarus, Russia and the Ukraine to Israel. Immigrants from these countries formed the association 'SOS Chernobyl'. According to this association 100.000 'victims of Chernobyl' arrived in Israel between 1990 and 1992. They were dissatisfied by what they considered a lack of concern for their medical problems by Israel authorities (Chernobyl conference, 1992, p.21). However, it is not clear what role the Chernobyl catastrophe really played in their decision to migrate to Israel. For that reason different estimates can be given of numbers of environmental refugees in this as well as in other cases.

A quite different indirect socio-economic effect may be caused by the investments in environmental protection measures by enterprises. Industries in countries without strong environmental control measures have to invest less in environmental protection measures than industries that have to conform to the stricter environmental policy standards of their countries. The latter can either plea for tariff barriers for products of polluting industries abroad or replace their activities to countries with less strict environmental regulations. Both developments can become a source of national or international tensions (See also section 10.9.3)

The provision of domestic consumer needs by exhausting natural resources including flora and fauna elsewhere can also be considered as a spatial transfer of environmental problems. Domestic environmental improvements can go hand in hand with on-going exploitation of the environment elsewhere. Japan is a country that combined a strict domestic environmental policy with extreme exploitation of environmental resources elsewhere. Its domestic environmental policy belongs to the most advanced in the world (Tsuru and Weidner, 1985; Maull, 1992). Its reputation as regards international environmental measures is far more negative. The country is one of the largest consumers of tropical timber of the world. No matter if trade in ivory or catching of whales was on the agenda Japan belonged to the camp of adversaries of restrictive measures and continued to commit heavy assaults on the environment abroad. Only as a result of strong international pressure has it given up resistance to international environmental protection measures in later years (Maull, 1992)

Another example of the going together of an ambitious national environmental policy and devastating environmental effects abroad is offered by the Netherlands. The more than 14 millions pigs (see figure 1.6) and other life-stock living in the Netherlands are fed with agricultural products from all over the world. 'For every acre of arable land in the Netherlands (2 million hectares, 0.13 hectare per inhabitant) seven more are cultivated abroad (of which three in developing countries: 6 million hectares) for the needs of the Dutch livestock industry and Dutch consumption and agri-exports' (IUCN, 1994, p.59). Tapioca is mainly imported from Thailand, soybean products from Brazil and Argentina and oil-palm products from Malaysia. The production of these agricultural products for the Dutch bio-industry contributes to the disappearance of tropical forests and soil erosion in the exporting countries.

It is reasonable to expect that direct and indirect transboundary effects of environmental damage will become major sources of international environmental conflicts in the near future. Will the period of the cold war be followed by a period of many green wars?

CIA turns green

What are the US soy satellites now looking at? The world's natural resources. Richard Norton-Taylor investigates the Agency's new fears of eco-wars (...) it is now a shared belief among the west's security and intelligence agencies that future conflicts increasingly will have their roots in the environment, a term which until not so long ago was marginalised as a largely apolitical concept. Conflicts over resources -land, food, water - will threaten the existing order and exacerbate simmering tensions between rich and poor, north and the south. The rich, including Europe, are already raising their drawbridges and targeting immigrants and asylum seekers. They are attacking the symptoms rather than the causes.

The Guardian, September 11, 1996

3.6 Sustainable development

Transfer of environmental problems also has a temporal dimension. In the second half of the twentieth century natural conditions and natural resources like minerals and fossil fuels that have existed or have been formed during millions of years are destroyed in a few decades. People living in this short era benefit from excessive use of the earth's scarce resources. Future generations will suffer from the exploitation and pollution of the earth by their predecessors at the end of the second millenium.

In 1987 the report *Our common future* of the World Commission on Environment and Development' (the Brundlandt commission; WCED, 1987) was published. The report intends to combine developmental and environmental policy on a world-wide scale. The key concept of the report is *sustainable development.* This concept has been defined as 'development that meets the needs of the present without compromising the ability of future generations to meet their own needs' (WCED, 1987, p.43).

The concept of 'sustainable development' has been interpreted in quite different ways. A distinction has been made between 'weak' and 'strong' sustainability (Opschoor, 1996). 'Weak' sustainability allows for dwindling of natural resources as long as they are replaced by human made products and human developed technologies. The criteria of 'weak sustainability' are fulfilled as long as synthetic materials can take over the functions of natural resources and as far as reduced availability of resources like fossil fuels can be compensated by more efficient use. 'Strong' sustainability requires the maintenance of quantities and qualities of natural resources without allowing for substitution by man-made resources. The distinction between 'weak' and 'strong' sustainability is strongly related to the difference between an eco-centric and an anthropocentric approach of nature and environment which will be discussed in section 4.9.1. Only from an anthropocentric viewpoint is a 'weak' definition of sustainability acceptable. It cannot be denied that the Brundtland definition of sustainable development implies an anthropocentric approach and is far removed from an eco-centric approach of the use of natural resources.

The concept of 'sustainable development' is sometimes considered as being the equivalent of 'stable development'. The Russians make this mistake when they translate 'sustainable' with their adjective 'ustoichivyi' which means 'stable'. Sustainable development has not necessarily to come into conflict with turbulent development, as long as environmental qualities are maintained for the benefit of future generations. Another mistake is to confuse the idea of 'sustainable development' with the ideal of a 'sustainable society'. Societies are always changing and only in utopias there are sustainable societies. The ambition of the Brundtland report and its adherents is not to end social change but to test and to adapt social change to the requirements of equal possibilities of environmental utilisation for future generations.

Chapter 4
Culture and civilization

4.1 Historical roots and contemporary dilemmas

Why is it that in the western world a type of society developed in which control and exploitation of nature reached levels that were never before attained in human history? Often 'western culture' is mentioned as a major cause of the unique developments that started in the west.

What is 'culture'? Many definitions of the concept of culture have been given. We will define culture as 'the transferable result of human efforts'. These efforts can be either material or abstract. Products and buildings are examples of material culture. Ideas and fantasies belong to the abstract culture. Abstract culture should be distinguished from purely individual ideas or fantasies. Abstract culture is always more than the exclusive property of individuals. It can be shared by and transferred to others.

It hardly needs to be explained that environmental problems are related to the use of specific forms of material culture like products and production processes ('technology'). However, environmental problems are also related to abstract culture, to the ideas of people about their relationships with nature in particular periods and in particular regions. This chapter deals with the environmental impact of abstract culture, in particular western culture. Among the elements of culture that will be discussed are religion, philosophy, political ideologies and science.

Culture can stimulate both human respect for and human exploitation of nature. Human respect for nature can be expressed in the form of a taboo. The word 'taboo' has been borrowed from Polynesian languages. It 'expresses a connection with gods, or a separation from ordinary purposes, and exclusive appropriation to persons and things considered sacred' (Van de Braak, 1995, p.27). By means of a 'taboo' human influence on living or non-living elements of nature can be restricted or even completely banned.

Modern forms of control and exploitation of nature could only develop after overcoming taboos. This can be said both of nature surrounding human beings and of human beings as part of nature. 'Just as opening the human body by the anatomist should originally be understood as the transgression of a sacred, dangerous boundary, the same applied in ancient times to the opening of the earth by a miner' (Van de Braak, 1995, p.51).

Not only restrictions of human impact on nature but also the opposite of it, human exploitation of nature can be stimulated by abstract culture. In the following section we will deal with the role which is attributed to Christianity and Enlightenment in the domination and exploitation of nature in the western world. We will then also discuss cultural sources of respect for nature and resistance against modern forms of exploitation of nature.

The concept of 'culture' is closely related to the concept of 'civilization'. The latter concept refers to the combination of both abstract and material culture in a particular period and/or era. It is strongly associated with the notion of control. It can refer to control of nature and of human behaviour. Therefore civilization may stimulate highly developed exploitation of nature for human purposes as well as inhibition of forms of human behaviour that might damage nature. In this chapter we will discuss processes of civilization and their environmental impact and contrast the ideas of those environmentalists who believe in on-going civilization and those who hope for a 'return to nature' in order to protect nature.

Nowadays we are confronted with a great cultural variety as regards the interaction between man and nature. People may differ widely in their opinion about the type of nature that should be protected and the types of environmental risks that are threatening human beings. After discussing these actual controversies we will conclude this chapter with a presentation of some continuing controversies within western culture in which the environmental issue plays a major role.

4.2 The domination of nature

4.2.1 Alienation: when did it start?

In critical observations on 'modern western culture' the origin of today's evils is attributed to different historical times or even to mythical events. The economic boom of the sixties, the industrial revolution, the era of 'enlightenment' and Eve's eating an apple of the Tree of Knowledge in the paradise are mentioned in this context. 'Alienation' is what happened in all these dramatic episodes. Modern man is considered as being alienated from the natural order no matter if this order is a paradisiacal unity with nature, a divinely sanctioned society of the Middle Ages, or an impoverished post World War Two society. In her historical study on environmental consciousness Anna Bramwell remarks: 'analysis of the seeds of anti-nature suffers from a bewildering variety of historical locations and cruces' (Bramwell, 1989, p.24). Nearly 2000 years old Christianity and slightly more than 200 years old Enlightenment are frequently presented as more or less related causes of the alienation of nature which led to our contemporary environmental crisis. (Gotlieb, 1996).

4.2.2 Christianity

It has often been argued that the Judeo-Christian culture removed God from nature. Whereas Eastern and Greco-Roman religions thought of both living and non-living nature as possessing a spiritual dimension, the God of Jews and Christians is not present in nature. 'In the Judeo-Christian doctrine, the distinction between the Creator and the created (even though proofs of the Creator's existence may, to some extent, be obtained from observing the creation) is unequivocal, as it must be: there can never be any question of the inferiority of the natural order, lovely as it is, to God' (Glacken, 1967, pp.196/197).

Genesis on the relationship between man and nature

Genesis 1. 26: Then God said, 'Let us make man in our image, in our likeness, and let them rule over the fish of the sea and the birds of the air, over the livestock, over all the earth, and over all the creatures that move along the ground.'
Genesis 1. 27: So God created man in his own image, in the image of God he created him; male and female he created them.
Genesis 1. 28: God blessed them and said to them, 'Be fruitful and increase in number; fill the earth and subdue it. Rule over the fish of the sea and over every living creature that moves on the ground'.
In another part of Genesis the idea of man subduing earth is formulated even more powerfully:
Genesis 9. 2: The fear and dread of you will fall upon all the beasts of the earth and all the birds of the air, upon every creature that moves along the ground, and upon all the fish of the sea; they are given into your hands.

Holy Bible

Often mentioned in this context is the story of the Creation in Genesis, the first book of the bible. There God is the distant creator of nature and man a creature that subjects all other living and non-living nature to his will. 'Fill the earth and subdue it' in Genesis 1.28 is a perfect guideline for destruction of nature by man. In a classic, though controversial, article Lynn White called the Judeo-Christian vision on the relationship between God, man and nature the basic cause of our contemporary ecological crisis (White, 1967). Apart from the domination of man over nature (as formulated in the bible book Genesis) he mentions two other, although related, elements of the Christian worldview, which have had a far reaching environmental impact:

– The belief in progress

The Judeo-Christian concept of time as non-repetitive and linear was in sharp contrast with the notion of time being cyclical which had prevailed until then. 'Our daily habits of action, for example, are dominated by an implicit faith in perpetual progress which was unknown either to Greco-Roman antiquity or to the Orient. It is rooted in, and is indefensible apart from, Judeo-Christian teleology' (White, 1974, p.23).

– God's transcendence of nature

'In Antiquity every tree, every spring, every stream, every hill had its own *genius loci*, its guardian spirit. These spirits were accessible to men, but were very unlike men; centaurs, fauns, and mermaids show their ambivalence. Before one cut a tree, mined a mountain, or dammed a brook, it was important to placate the spirit in charge of that particular situation, and to keep it placated. By destroying pagan animism, Christianity made it possible to exploit nature in a mood of indifference to the feelings of natural objects' (White, 1974, pp.24/25).

Christianity as an ideology of anti-environmentalists

We anti-environmentalists hold to the Judeo-Christian view that man is emphatically not part of nature. We hold that human life has a different kind of value from animal life: that protecting and preserving human life is a moral duty that sweeps away all 'duties' to nature, and the very idea of 'duty to nature'.

The Washington Post, November 17, 1996

White's thesis has been opposed with different arguments. His interpretation of the bible has been criticized. Even the text above quoted from Genesis 1:26-28 could be interpreted as a call for management instead of exploitation of nature. Christianity is a pluriform religion and it also includes the contrast between the distant and subduing approaches to nature. White himself mentions one such contrasting view which is represented by St.Francis of Assisi, who preached humility of man towards both living and non-living nature. For him 'Brother Ant and Sister Fire (are) praising the Creator in their own way as Brother Man does in his' (White, 1974, p.29).

Between these extremes of dominance and humility Christianity also justifies an attitude towards nature in which recognition of the power of man over (other parts of) nature is combined with the conviction that this power implies responsibility for the well-being of nature. This vision of the relation between man and nature has been called stewardship. A justification for this vision can be found in Genesis 2.15 where it is said that 'The lord took the man and put him in the Garden of Eden to work it and to take care of it'. The life and work of Benedict of Nursia, who founded a monastery in the sixth century, have been considered as exemplary for the attitude of stewardship. 'Throughout the history of the Benedictine order, its monks have actively intervened in nature - as farmers, builders, and scholars. They have brought about profound transformations of soil, water, fauna, and flora, but in such a wise manner that their management of nature has proved compatible in most cases with the maintenance of environmental quality' (Dubos, 1974, p.132).

At least three different Christian attitudes towards nature can be distinguished:

- the domination of nature by man as stated in several passages of Genesis;
- the humility towards nature of Francis and his followers;
- the stewardship of nature practised by Benedictine monks and sisters.

Besides that there are differences between Roman and Byzantian Christianity, between Roman Catholicism and Protestantism and between different denominations of these forms of Christianity. According to some authors the Christian belief in the domination of nature by man should be ascribed primarily to Greek instead of Jewish influences on Christianity (Passmore, 1974, p.17).

It is striking that in the Middle Ages, a period in which Christianity dominated individual and social life, there existed strong religious resistance towards all sorts of human intervention in nature. Nature as created by God was considered as a holy order which should not be changed by human intervention. In his voluminous study on the appreciation of technical progress Van der Pot mentions different examples in the Middle Ages of resistance to human intervention in nature based on religion:

- Dying of clothing was rejected with the argument that God would have created sheep of different colours if he had appreciated people dressing in different colours.
- Mining of metals was opposed because if God had wanted metals available for human use he would not have put them deep down in the earth.
- In Spain, during the reign of king Philip IV (1605 - 1665), a plan to make rivers navigable was rejected with the argument that man has no right to end the work that God, for incomprehensible reasons, did not finish (Van der Pot, 1985, pp.95-96).

Even today forms of Christian inspired resistance to human intervention in nature exist, in particular as far as human life itself is seen to be part of nature. In western countries Christians oppose legalisation of human intervention at the beginning of life (abortion) and at the end of it (euthanasia). A comparatively new form of human interference in both animal and human nature with unknown but probably far reaching consequences is genetic engineering. Again this is opposed by some Christians. So, Christianity both justifies exploitation of nature by man and also restricts human manipulation of nature.

It has been argued that the domination of nature by man, as formulated in Genesis, is typical of all forms of monotheism and not only for Christianity (Toynbee, 1974). If this were true, the damaging environmental effects which White ascribes to Christianity could also be ascribed to Islam. Like the Bible, the Koran stresses both man's mastery over and his stewardship of nature (Baird Callicott, 1994, p.31)

We discussed above different Christian attitudes towards the relationships between man and nature. However, the question of whether and to what degree contemporary environmental problems are caused by Christianity has not yet been answered. Several authors who contributed to the debate on the postulated Christian roots of the contemporary environmental crisis argue that religious convictions do not have any or at most a marginal influence on human behaviour towards nature. They indicate that environmental destruction took place and still takes place in societies with widely different cultures. In our chapter on history we already have mentioned several examples of pre-Christian environmental destruction.

Environmental damage caused by human beings also takes place in parts of the world which were hardly or not at all influenced by Judeo-Christian culture. Moncrief (1974) uses this as an argument against White. He ascribes forms of environmental degradation primarily to factors like urbanization, increased wealth, increased population and individual ownership of resources (Moncrief, 1974, p.90). According to this author Christianity only had any influence on the human use of nature indirectly and via the above mentioned factors.

No definitive statements about causal relationships between Christianity or any other religion on the one hand and environmental decay on the other hand can be made. However, it is clear that according to Christianity God is not part of nature but the distant creator of it. Unlike animism Christianity implies the belief in a nature free of spirits. It does not recognize spirits and taboos which prevent human manipulation of nature. Therefore Christian and post-Christian societies need other forms of restriction of human influences on nature.

4.2.3 Enlightenment

In European cultural history the period between the sixteenth and eighteenth century is known as the period of 'enlightenment'. The philosophy of René Descartes (1596 - 1650) introduced a fundamental change in scientific thinking. He rejected the Aristotelian based scientific practice of his age, which in his opinion consisted too much of reformulation and reordering existing knowledge and insights. Descartes appreciated mathematics highly and was astonished to discover that it was only applied in the 'mechanical arts'(Descartes, 1965, p.8). It was his idea that mathematics should be used to describe all natural phenomena. He argued for a study of living nature in which complicated forms of life are explained by the properties of their smallest component elements. The philosophy of Descartes contributed to what was later called 'the mechanization of the world picture' (Dijksterhuis, 1969) and to the experimental manipulation of nature. According to Descartes scientific knowledge should contribute to 'making ourselves the masters and possessors, as it where, of nature' (Descartes, 1965, p.50). In Britain the same ambition to develop scientific knowledge in order to master and to manipulate nature was defended by Francis Bacon (1561 -1626). In his utopian sketch of 'New Atlantis' he summarized his ambitions as follows:'The end of our foundation is the knowledge of causes and secret motions of things,

and the enlarging of the bounds of human empire, to the effecting of all things possible.' (Bacon, no date, p.33).
In the concept of 'Mother Nature', the idea of motherhood and the idea of nature are combined.

The domination of nature and women

In her study on the replacement of an 'organic' by a 'mechanical' view of nature during the era of enlightenment Carolyn Merchant gives much attention to the identification of nature with women: 'Central to the organic theory was the identification of nature, especially the earth, with a nurturing mother: a kindly beneficent female who provided for the needs of mankind in an ordered, planned universe. But another opposing image of nature as female was also prevalent: wild and uncontrollable nature that could render violence, storms, droughts, and general chaos. Both were identified with the female sex and were projections of human perceptions onto the external world. The metaphor of the earth as a nurturing mother was gradually to vanish as a dominant image as the Scientific Revolution proceeded to mechanize and to rationalize the world view. The second image, nature as disorder, called forth an important modern idea, that of power over nature'

Merchant, 1983, p.2.

The philosophy of the enlightenment paved the way for a ruthless exploitation of nature, including animals (Noske, 1989). However, as in the case of Christianity, one should be careful not to give a one-sided impression of the ideas of this era. The era of enlightenment was also the period in which the famous French philosopher Jean-Jacques Rousseau (1712-1778) wrote his works in which he formulated his belief in the goodness of nature and rejected the evils of civilization. Rousseau came to inspire widely different ideological forms of protest against dominant developments within western culture (see section 4.3). Modern science also stimulated respect for nature as we shall see in section 4.4.

Rousseau on man and nature

Everything is good as it leaves the hand of the Author of things; everything degenerates in the hands of man. He forces one soil to nourish the products of another, one tree to bear the fruit of another. He mixes and confuses the climates, the elements, the seasons. He mutilates his dog, his horse, his slave. He turns everything upside down, he disfigures everything; he loves deformity, monsters. He wants nothing as nature made it, not even man; for him, man must be trained like a school horse; man must be fashioned in keeping with his fancy like a tree in his garden.

Rousseau, 1979, p.37.

4.3 Anti-capitalist ideologies

4.3.1 Industrial capitalist society and its opponents

Modern western societies as they have developed since the age of enlightenment, differ from all earlier types of societies by their great technological ability to control and to exploit nature. Socially these societies can be regarded as highly differentiated: different human activities take place in different social unities. Work and private life are institutionally separated, but individuals and their families are highly dependent on having jobs, not only to receive money but also to be respected as valuable citizens.

The type of society which is called 'modern' here, did not develop without violence and repression. According to Barrington Moore 'all forms of industrialization so far have been revolutions from above, the work of a ruthless minority' (Moore, 1966, p.506). Dickson, who presents several cases in which technological innovation was used as a means of repression, also mentions the struggle of the Luddites to destroy machines at the beginning of the nineteenth century. The government force of 12,000 men, fighting the Luddites in England in the period 1811-1813, was larger than Wellington's army, which was contributing to the defeat of Napoleon on the continent in the same period (Dickson, 1974, pp.77/78).

In the nineteenth and twentieth century, different political movements opposed the liberal, industrial-capitalist society of their day. The environmental damage caused by that type of society played a role in this opposition. We will discuss four of these movements here, which differ not only in their basic ideas but also in the degree to which they were institutionalized in political parties or political systems. These four are: communism, anarchism, conservatism and fascism.

In our so-called 'post-modern' era, these anti-capitalist movements have lost the significance which they had in the past. But they represent different critical reactions to modern societies and their environmental effects which have not disappeared today and which may be heard even louder in a future post-postmodern era.

4.3.2 Communism

Communism as the ideal of social equality among people has a long history. The main founding father of 19th and 20th century communism was the German Karl Marx (1818-1883). Marx criticized the capitalism of his day, an economic system in which the means of production are privately owned by capitalist shareholders and in which the workers are obliged to sell their working power to the capitalists in order to earn enough money to survive. In this economic system workers are *alienated* from their own means of production and from the products which result from their productive efforts. Labour does not satisfy the need of products among the workers and of society but only serves the financial interests of the capitalists. Capitalism is an irrational system of production in which periods of over-production and wastage alternate with periods of stagnation and mass unemployment. It was Marx's idea that this system would come to an end when the growing proletariat of alienated labourers would take over the power of the shrinking class of capitalists.

Marx was convinced of the extremely important place of labour in human life. Labour enables the worker to create and maintain relationships with nature. Therefore capitalist alienation of labour also implies alienation from nature.

Marx on alienation of nature and work

In modern agriculture, as in urban industry, the increase in the productivity and the mobility of labour is purchased at the cost of laying waste and debilitating labour-power itself. Moreover, all progress in capitalist agriculture is a progress in the art, not only of robbing the worker, but of robbing the soil; all progress in increasing the fertility of the soil for a given time is a progress towards ruining the more longlasting sources of that fertility. The more a country proceeds from large-scale industry as the background of its development, as in the case of the United States, the more rapid is this process of destruction. Capitalist production, therefore, only develops the techniques and the degree of combination of the social process of production by simultaneously undermining the original sources of all wealth - the soil and the worker.

Marx, 1976, p.638.

Marx distinguished between 'relations of production' and 'forces of production'. The first term refers to the (capitalist) ownership of means of production. The latter to the technology and organization of (industrial) production. Marx rejected the first, but not the latter. He thought that further development of industry would be stimulated by the abolition of private ownership of the means of production and that the entire society, instead of just a minority of shareholders, would profit from this change in the relations of production. The consequences of the absence of fundamental criticism of the industrial forces of production became visible only after his death.

The term 'Marxism-Leninism' is used to refer to the type of communism which developed after the Russian October revolution of 1917. This communist revolution was a reaction to the preceding 'bourgeois' February revolution which had already abolished Tsarist power. For that reason, the October revolution can be considered as a counter-revolution against the development of a liberal-democratic society.

Marxism-Leninism can be qualified as an extreme ideological support for the benefits of industrial production. In this respect it is a typical example of western belief in progress by means of exploitation of nature.

During the Stalinist period 'nature' was considered as an obstacle to further progress (Weiner, 1984, p.75). Scientific disciplines like genetics and ecology, which showed either the limited opportunities available to humans for changing nature or the damaging effects which human activities could have on the natural environment, were condemned as bourgeois and counter-revolutionary. Scientists working in this field, particularly in genetics, were dismissed, arrested and liquidated. Large projects were started, aimed at transforming or correcting what were called 'mistakes of nature'. The term **voluntarism** has been often used to refer to the arbitrary and frequently impulsive manner in which sweeping changes in natural conditions were pursued, without any thought being given to the real possibilities for change and the damaging impact of such changes. One example of such a large-scale project is the one which was designed to change the direction of flow of certain stretches of rivers in Siberia and the European part of Russia from north to south. The plan was officially abandoned in 1986.
Marxism-Leninism became an ideological justification for extreme aggressiveness and carelessness towards nature, and a driving force behind some of the most extreme forms of environmental destruction committed by human beings. On the other hand, Marx's criticism of the irrationality of capitalist production is of continuing importance. Until now market economies have not succeeded in developing an economic system which both satisfies need for work of the (potential) working population and allows sustainable use of natural resources. (See section 3.2.1.).

4.3.3 Anarchism

As early as the nineteenth century Marxism was opposed by anarchism because of its authoritarianism and striving for power. The idea of an anarchist organization is something of a contradiction in terms. Anarchists highly praise freedom and spontaneity and are less charmed by organizations and their built-in tendency toward hierarchy and bureaucracy. For that reason, and in contrast to communists, anarchists did not develop long-lasting political organizations operating on a world-wide scale.
Whereas liberal-capitalist society is based on competition, anarchists consider cooperation as the main force needed to reach their goals. In his study 'Mutual aid' the famous Russian anarchist Peter Kropotkin (1842 -1921) argued that cooperation is a major force in the evolution of both animals and human beings. A characteristic of anarchists is their optimisms regarding free and independent citizens who are not hindered by authoritarian control from above. It is their idea that non-authoritarian communities will be able to develop sustainable relationships with their surrounding natural environment.
Unlike Marxists, anarchists are opposed to the industrial 'forces of production' which developed under capitalism. Anarchists argue that in modern industrial societies people suffer from lack of individual freedom and social equality. They

strive for non-hierarchical communities of free people.

Exploitation of man and nature

The basic conception that humanity must dominate and exploit nature stems from the domination and exploitation of man by man. Indeed, this conception goes back earlier to a time when men began to dominate and exploit women in the patriarchal family. From that point onward, human beings were increasingly regarded as mere resources, as objects instead of subjects. The hierarchies, classes, propertied forms, and statist institutions that emerged with social domination were carried over conceptually into humanity's relationship with nature. Nature too became increasingly regarded as a mere resource, an object, a raw material to be exploited as ruthlessly as slaves on a latifundium. This 'worldview' permeated not only the official culture of hierarchical society; it became the way in which slaves, serfs, industrial workers and women of all social classes began to view themselves. As embodied in the 'work ethic', in a morality based on denial and renunciation, in a mode of behaviour based on the sublimation of erotic desires, and in other worldly outlooks (be they European and Asian), the slaves, serfs, workers, and female half of humanity were taught to police themselves, to fashion their own chains, to close the doors on their own prison cells.

Bookchin, 1980, p.40-41.

Anarchism can be considered as a manifestation of the old longing for a 'return to nature', a return to a both technically and socially simpler way of life. There was a great, renewed interest in anarchism in western countries during the sixties and seventies of the twentieth century. Among the influential writers of that period are Paul Goodman (1970) and Murray Bookchin (1980, 1982). The environmental movement of that time was strongly inspired by anarchists ideas. The belief that 'small is beautiful' (Schumacher, 1973), the preference for small organizations and simple technologies among environmentalists and the high value attributed to voluntary cooperation were closely related to old and new anarchist ideas.

The least that can be said about the actual meaning of anarchism is that it is a correction to a one-sided fixation on the market and/or the State as the only institutions which are able to limit environmental destruction by man. Environmentally friendly alternatives, in particular in the field of the production and distribution of food and the use of renewable sources of energy, have originally been initiated as voluntary activities outside the field of market regulated or government regulated activities.

However, from the viewpoint of environmental protection, anarchism also has considerable weaknesses. First of all, small is not always beautiful. In reality small enterprises are, if compared on the basis of a specified product output, often more polluting than large ones, because the latter are better able to introduce and handle environmental control devices and measures. The same can be said of the preference among some anarchists for life in the country-side instead of in cities. Citizens in general need less energy to heat their houses and less car kilometres to move to other places than people living in the countryside (Paehlke, 1989, p.245).

Another weakness of anarchism is its lack of understanding of the role of countervailing power (see section 3.2.2.). The free people of anarchist communes do not always spontaneously respect their fellow man and their natural surroundings. The latent presence of outside control can help to prevent social and environmental repression.

A last point of criticism is that anarchism suffers from a lack of understanding of the role of complicated technologies and organization in the striving for a sustainable use of natural resources. Because of the numbers and the needs of human beings living on earth today, a modest and environmentally adapted use of scarce environmental resources will not be attained without the development of modern, complicated technology and organizations.

4.3.4 Conservatism

'Conservation' and 'conservatism' are not only linguistically closely related. What is more 'conservative' than the protection of existing ecosystems and landscapes? Yet environmentalists are not always conservatives in the socio-political sense. In the context of the latter meaning, it is possible to distinguish a 'conservative' approach to environmental problems from other approaches. There are conservative political parties but 'conservatism' as an ideology is not the monopoly of political parties. It is a vision on social issues which is far more widely spread.

Collapse of society

The developed nations consume such disproportionate amounts of protein, raw materials and fuels that unless they considerably reduce their consumption there is no hope of the undeveloped nations markedly improving their standards of living. This vast differential is a cause of much and growing discontent, made worse by our attempts at cultural uniformity on behalf of an expanding market economy. In the end, we are altering people's aspirations without providing the means for them to be satisfied. In the rush to industrialize we break up communities, so that the controls which formerly regulated behaviour are destroyed before alternatives can be provided. Urban drift is one result of this process, with a consequent rise in anti-social practices, crime, delinquency, and so on, which are so costly for society in terms both of money and of well-being.

Ecologist, 1972, pp.23/25

For conservatives environmental destruction is related to the disturbance of traditional forms of social life like families and local communities. They criticize modern society both for its damage to nature and landscape and for its destruction of traditional social bonds, resulting in all sorts of so-called 'deviant behaviour'.
It is usual to consider communism and anarchism as 'left-wing' and conservatism and fascism as 'right-wing' ideologies. The base of such a classification is the difference in vision on social equality. Communism and anarchism both strive, though in quite different ways, toward social equality whereas conservatism and fascism can be considered as ideological justifications of the maintenance of social inequality and the restoration of hierarchy.
This difference is quite clear if we compare anarchism and conservatism. Both are critical of the development of modern industrial capitalist society. However, anarchists criticize the absence of freedom and equality in modern societies, whereas conservatives criticize the lack of hierarchy and the prevalence of all sorts of 'deviant behaviour' in modern societies. Anarchists love spontaneity, conservatives hate lawlessness. Yet there are also socio-political writings in which conservative and anarchist visions are mixed. Interestingly enough the anarchist author Paul Goodman mentioned above presented himself as a conservative near the end of his life (Goodman, 1970). The 'Blueprint for survival' also offers a mixture of conservative and anarchist ideas on the social causes and social solutions of the contemporary ecological crisis (Ecologist, 1972). With anarchists they share a plea for small and self-regulating groups in order to avoid bureaucratic top-down regulation, and together with conservatives they stress the importance of hierarchy and social control not only to prevent environmental disruption but also all sorts of deviant behaviour like crime, illegitimacy, alcoholism, mental disease and suicide.

4.3.5 Fascism

In theory fascism on the one hand and anarchism and conservatism on the other hand have elements in common. In practice there were striking parallels between societies developed according to the ideas of Marxism-Leninism and those based on the ideas of (German) national-socialism and (Italian) fascism. Hereafter the term fascism is used in a broad sense referring both to the German and Italian variant. Originally, Italian fascism in particular was supported by numerous idealistic intellectuals and artists (Hamilton, 1973). At that time fascism appealed to people who disliked 'mass society' as it had developed under the influence of industrial capitalism. There is some similarity between the belief in individual freedom and spontaneity among anarchists and the glorification of 'living dangerously' among Italian fascists. There is a certain relationship between the conservative protection of traditional life and the fascist protection of unspoiled 'blood and soil' as well. Fascism can be considered as a revolt against the complexities of modern industrial societies and its differentiated social life. With its strong craving for a simpler life it shows parallels with both anarchism and conservatism. Between the two world wars among the young German 'Wandervögel' there were sympathies for forms of anarchy and communism as well as for racially pure settlements or communes (Bramwell, 1989, p.189). Among national socialists there was a strong identification with German nature, and in particular German forests.

Forest protection in Nazi-Germany

Arguably, no German government had ever taken the protection of the German forests more seriously than the Third *Reich* and its *Reichsforstminister* Göring. Reluctant eleven-years-olds were turned into expert leaf-peepers through programs on forest ecology introduced to schools and were shown how the woodlands demonstrated the laws of biological competition and survival from the earwig to the eagle. Conservation was institutionalized through the creation of an entire administration run by the likes of Schönichen who lectured on the subject at Berlin University from 1934 to 1936.

Schama, 1995, p.119.

The fascist variant of the idea of the goodness of nature was the plea for pure 'blood-and-soil', justifying the destruction of Jewish and other foreign 'blood' on German 'soil'.

For several decades after World War Two, people in western industrial capitalist societies could live with the illusion that fascism was a phenomenon of the past. However, there has been a revival of neo-fascist racism in Europe over several years leading to great electoral support to neo-fascist political parties. Other phenomena that may be considered as manifestations of such a revival are aggression towards immigrants in different countries and 'ethnic cleansing' in the former Yugoslavia. This neo-fascism, this struggle for the protection of one's own blood and soil against 'foreign elements' could in theory become related to strong pleas for environmental protection. It is not at all difficult to integrate the striving for pure blood and soil with the struggle for an unspoiled and pure nature.

4.3.6 Contemporary relevance of past ideologies

Communism, anarchism, conservatism and fascism are more or less ideologies of the past. However, they all contain elements which in a quite different context might return in future forms of environmental protest. Like communists, environmentalists of the future may protest against the irrationalities of production in market economies and argue for stronger control by society on decisions about production, possibly on a world-wide scale. Like anarchists they may experiment with small-scale technology and organization in order to develop an environmentally adapted way of life. Like conservatives of the past they may defend their local environment and society against destructive influences from outside. Like fascists they may combine resistance to destructive exploitation of nature with opposition to other 'unnatural elements' like 'foreign' ethnic groups and lifestyles. For that reason past experiences with anti-capitalist ideologies are of continuing interest for those who want to understand contemporary and future reactions against human-induced environmental change.

4.4 Science and respect for nature

From the leading authors of the age of enlightenment one may get the impression that the development of modern science only led to further human dominance and exploitation of nature. This impression is one-sided and incomplete. The British author Keith Thomas offers a totally different image of developments in Britain between 1500 and 1800. He argues that modern science of this period with its systematic observation and classification of observed phenomena contributed to a less anthropocentric view on non-human nature. In the past the animal world was ordered by means of the categories edible and inedible; wild and tame; useful and useless (Thomas, 1984, p.53). The scientific study of nature introduced classifications of nature which were not related to such practical human interests. 'Henceforth, the systematic investigation of nature would be conducted on the assumption that plants and animals should be studied for their own sake, independent of their utility or meaning for man' (Thomas, 1984, p.91). In the nineteenth century it was Charles Darwin (1809-1882) with his theory of evolution who further undermined the Judeo-Christian belief in the uniqueness of the human species as it was formulated in the book of Genesis. Thomas does not deny the environmental impacts of modern science and technology. In his opinion, modern science stimulated both domination of nature and respect for nature. The conclusion of our earlier remarks about the age of enlightenment and of this section is that on the one hand modern science enables excessive control and exploitation of nature and for that reason can be considered as one of the main causes of contemporary environmental problems. On the other hand it inspires respect for and protection of nature.

The famous statement of Descartes 'cogito ergo sum' (I think, therefore I am) was the expression of extreme detachment from nature. Reflection as the base of science was completely detached, not only from nature surrounding the thinking scientist but even from his own physical existence. In many studies this detachment is considered as a major cause of excessive forms of human exploitation of nature. Sheldrake (1993) argues for a less detached attitude towards nature and the recognition that human beings are part of nature. However, not only 'closeness to nature' but also 'detachment from nature' can inspire nature conservation and environmental protection. After discussing the phenomenon of civilization and civilization processes we will return to this dichotomy of closeness to and distance from nature.

4.5 Civilizing processes

4.5.1 Civilization and control

Development of respect for nature, as described above by Thomas, can be considered as an indication of increasing civilization. The concept of civilization refers to both abstract and material culture. It has often been used to refer to all things in which western culture is considered to be superior to other cultures. Nobert Elias, whose work inspired the writing of this section, wrote the following about the term 'civilization': 'It sums up everything in which western society of the last two or three centuries believes itself superior to earlier societies or 'more primitive' contemporary ones. By this term western society seeks to describe what constitutes its special character and what it is proud of: the level of *its* technology, the nature of *its* manners, the development of *its* scientific knowledge or view of the world, and much more.' (Elias, 1978, pp.3/4). However, there is another common element in the phenomena which, according to Elias, are part of the process of 'civilization'. They all have to do with control: self-control, control by society and control of nature and the environment.

In his study of the process of western civilization since the Middle Ages, Elias related changes in individual behaviour and sentiments to changes in society. Table manners and types of behaviour regarding other physical functions that were normal in the Middle Ages, were no longer considered as respectable after that period. People became more restricted in their behaviour and developed feelings of uneasiness and shame in connection with forms of behaviour which were related to physical functions. Elias explained this process of changing behaviour and sentiments as resulting from processes of centralization, state formation and state monopolization of force that took place in the same period. Later Elias included control of nature and the environment as a third element in his 'triad of control' consisting of non-human events (occurring in nature or environment), interhuman events and intrahuman events. This third form of civilization was extensively studied in Goudsblom's study on 'Fire and civilization' (Goudsblom, 1992).

What can be said about possible environmental effects of the supposed growing control or growing 'civilization' of intrahuman, interhuman and non-human events in western society?

4.5.2 Control of intrahuman events

Self-control or avoidance of forms of behaviour such as dumping waste in canals, excessive use of electric light or travelling by car in cases where public transport is available contribute to environmental protection. In chapter 6 we will discuss environmental attitudes and their relationships to environmentally relevant behaviour. In recent years a greater sensibility towards several forms of environmental degradation have been developed and feelings of guilt and shame became attached to types of behaviour that contribute to environmental decay. However, in general the relationship between attitudes and behaviour is weak. Economic conditions of individual citizens are more powerful stimuli for behavioural change than individual feelings of shame and guilt. Yet the maintenance and strengthening of intrahuman inhibition of forms of behaviour may contribute to the protection of the environment and therefore environmental quality may profit from progressive intrahuman civilization.

4.5.3 Control of interhuman events

Elias sketches an ideal of civilization in which well-behaved citizens have delegated the means of violence to the powerful modern state. In the case of fire, continuing civilization implies a growing dependency of individuals not only on the state, but also of other public and private institutions. Take the example of electricity: it offers individual consumers possibilities which surpass the wildest dreams of utopian thinkers of the past. However, as users of electricity, individuals are completely dependent on technologies and organizations which they hardly know, let alone can control.

Collectivization of motorized means of transport by public or private organizations and a reduction of the use of privately owned motorized means of transport in general will reduce the environmental damage and nuisance caused by traffic. Growing interhuman control and even monopolization of activities which have disturbing environmental effects can contribute to the improvement of environmental quality.

4.5.4 Control of environmental events

From the viewpoint of an environmentalist, civilization as a process of growing control over the physical environment is a mixed blessing. Take the example of fire (Goudsblom, 1992). On the one hand progress in the development of control can imply: the prevention or at least successful fighting of unwanted and destructive fires and more efficient and less polluting exploitation of energy. On the other hand progress in the development of control of nature also implies more use of natural resources and increased emission of pollutants. The process of civilization as described by Elias and Goudsblom may also result in environmental catastrophes. For example, nuclear energy could only develop under conditions of highly developed control of nature, self control of nuclear workers and state control of energy production. But these controls did not exclude the possibility of accidents like the one in Chernobyl in 1986.

From an environmental viewpoint progress in control of nature is a dubious benefit for at least two reasons. In the first place 'civilized' human activities can destroy nature on purpose, as in the case of the felling of tropical forests by multinational enterprises. In the second place they can have unintended 'external effects' (see chapter 5) both in the form of more or less permanent environmental stress and in the form of rare accidents. Modern environmental risks are an outcome of the civilization process. A civilized society, in the sense of Elias, is a 'risk society' (Beck, 1992). This concept will be discussed in chapter 7.

4.5.5 The ideal of a civil society

Especially since the disappearance of communist regimes in eastern Europe, the concept of 'civil society' has become popular. It expresses the ideal of a new society which should replace the former model of the authoritarian societies of central and eastern Europe. 'Civil society was seen as the opposite of despotism, a space in which social groups could exist and move - something which exemplified and would ensure softer, more tolerable conditions of existence' (Hall, 1995, p.1). A 'civil society' is a society in which citizens actively contribute to the well-being of their society. According to the Russian environmental sociologist Oleg Yanitsky one of the main challenges for the environmental movement in his country is to move from 'political mobilization' to (contributing to) 'civil society' (Yanitsky, 1993, pp.128-135). This concept of 'civil society' consisting of actively participating citizens is different from the concept discussed above of a civilized society. It is clear that environmental protection requires the contribution of environmental groups of independent and active citizens. We will discuss their organization in chapter 8.

4.6 Romantic and civilized environmentalism

Western culture, as it developed during the past centuries, has been characterized by contradictory developments of alienation from and return to nature. The development of modern science and technology enabled a growing technological control and exploitation of nature and human beings, inside and outside the processes of production. Time and again this process of civilization was opposed by pleas for a 'return to nature', implying greater respect for nature and for human spontaneity, or: a less detached attitude towards nature and a recognition of the unity between man and (other) nature. However, modern science does not only lead to a more perfect exploitation of nature but also stimulates respect for nature; modern technology does not only contribute to excessive exploitation of nature but is also a necessary prerequisite for a more efficient and less destructive human use of nature.

If we use the term 'environmentalism' as a common name for all forms of concern about environmental degradation caused by human activities we may distinguish between 'romantic' and 'civilized' environmentalism. These terms are used here as sociological concepts and not as value judgements.

Romantic environmentalists are representatives of the age old strive for a 'return to nature'. In their opinion men have become too alienated from nature. Modern civilization is considered as a barrier between man and nature. 'Foreign' influences disturb 'natural' relationships between men and their environment. The Franciscan variant of Christianity is an example of 'romantic environmentalism'. Of the ideologies discussed anarchism and fascism in particular contain elements of romantic environmentalism. A more recent form of romantic environmentalism is the so-called 'deep ecology' movement. This term 'deep ecology' was introduced in 1973 by the Norwegian philosopher Arne Naess and has become popular as the expression of radical criticism of the prevailing western way of thinking about man and the environment, and in particular of the idea of dominance of human beings over non-human nature (Devall and Sessions, 1985; Ecologist, 1988).

'Civilized' environmentalists are convinced of the necessity of further development of science and technology and further self-control of human behaviour in order to protect the planet earth. They do not agree with the idea that detachment from nature is an enemy of care for nature. 'Respect'is the term which best expresses the combination of detachment and care that is typical of what is called here 'civilized' environmentalism.

Nowadays 'ecological modernization' (section 3.2.1.) is a good example of what civilized environmentalists strive for, but they will add that in order to protect the environment, in addition to technical and organizational changes within the production process, more restricted or 'civilized' patterns of consumption are necessary as well.

Although it is useful to distinguish between civilized and romantic environmentalism as two widely different philosophies behind nature conservation and environmental protection, the contrast should not be exaggerated. What begins in the context of romantic environmentalism can later be continued within the frame of civilized environmentalism. The development of so-called alternative technologies offers examples of this transformation. In the Netherlands, modern use of wind energy was started by people who preferred a simple way of life close to nature in the countryside or along the water-side. Nowadays it is part of the electricity supply sector and considered as a technologically and organizationally advanced form of electricity production. The same story can be told about the development of the production and consumption of all sorts of 'eco' products.

We have given an impression of often contradictory and paradoxical historical developments within western culture and civilization and their environmental implications. At the end of this chapter we will discuss actual cultural controversies regarding nature conservation and environmental protection

4.7 Which nature should be protected?

'Free' and 'undisturbed' nature has often been contrasted with the 'man-made' environment. However, if only because of the enhanced greenhouse effect, 'nature' all over the world has been influenced by human behaviour and declared 'dead' for that reason (McKibben, 1989). 'Nature' in populated areas has been influenced by the past activities of human beings. This leads to a dilemma for those who love nature. Should plants, animals and ecosystems which developed under the influence of human activities in the past be protected or should the development of new nature, including the import of species of animals and plants from abroad, be stimulated?

We distinguished between 'conservative' and 'anarchist' opposition against modern society. In the former ideology, modern society is criticized because it undermines and destroys existing order. In the latter it is criticized because it represses the development of free people. The contrast between those who strive for 'nature conservation'and those who try to stimulate 'nature development' shows some resemblance to this contradiction. Those who argue for 'conservation' try to maintain 'their' nature under changing external circumstances, whereas those who strive for 'development' try to create possibilities for free development of 'new' nature. These two different approaches to nature conservation might be called conservative and anarchist nature conservation. Should nature as it has been formed in the past under the influence of human activities be protected or should development of new nature be given a chance? This controversy is not only a 'luxury' problem of prosperous western countries, but it also manifests itself in developing parts of the world, where the protection of existing species and ecosystems can hinder the development of sustainable use of natural areas.

Conservation versus sustainable use of nature in Africa

For rangeland managers of the 1960s and 1970s, domesticated livestock were an intrusive element that destabilised 'natural' botanic systems. The notion that domesticated herbivores were a foreign intrusion had considerable intuitive appeal in North America and Australia where European-owned stock had suddenly burst upon the scene, where the natural 'before' and the disturbed 'after'were clearly demarcated over historical time, and where the requirements of industrialised agriculture confronted a nostalgia for bygone landscape' (...) 'For agriculturists, the conservation of pristine vegetation is of less concern than the expected length of time that output can be maintained from altered vegetative states under different management regimes. In short, the objective of botanical immutability is less useful than a workable notion of sustainability.

Behnke and Kerven, 1994, pp.1-2.

Controversies about nature protection policies should not only be interpreted as controversies between 'the' interest of nature protection and other interests. They are also based on controversies among those who give top priority to nature protection.

4.8 Which environmental risks are threatening?

In this book we stress the fact that man-made environmental changes only become 'problems' when they are experienced as such by human beings. Between men, from laymen to scientific specialists, there are huge differences in the appreciation of human-induced environmental changes, varying from overall positive and optimistic to negative and pessimistic valuations. Some optimists argue that actual developments on earth do not at all justify alarmist statements about global environmental risks. Simon is a well-known environmental optimist. In a debate with Myers he praised the benefits of population growth and argued that 'almost every trend that affects human welfare points in a positive direction' (Simon, 1994, p.6). Examples are the continuous increase of life expectancy, the growing quantities of available raw materials and food and the decline of air pollution. Simon's view on so-called environmental risks can be summarized with a statement of Franklin Delano Roosevelt who once said that 'the only thing we have to fear is fear itself' (Myers and Simon, 1994, p.xiii).

As there are basic differences in opinion about the value of different types of nature there are also fundamental disagreements about the benefits and risks of human-induced environmental change. Cotgrove renamed environmental optimists and pessimists 'cornucopians' and 'catastrophists' and argued that this difference in attitude is related to having a job within or outside the industrial sector of society. According to this author 'environmentalism' is widespread among people in the non-productive sectors of society. It is a form of opposition of people on the periphery of society against the basic institutions of industrial capitalism and its dominant economic values (Cotgrove, 1982, p.95). 'The cornucopians are typically engineers, consultants, directors, managers, marketing and sales people, industrial scientists. The catastrophists are typically research scientists, academics, doctors, teachers, clergy, social workers, writers, actors and artists' (Cotgrove, quoted by Douglas and Wildavsky, 1982, p.162). This basic difference in the appreciation of human-induced environmental changes has been further elaborated in studies by Douglas and Wildavsky (1982) and Schwarz and Thompson (1990). The latter make a distinction between four myths of nature (see figure 4.1.) which they relate to different managerial attitudes.

The four myths of nature are described as follows:

'– *Nature benign* gives us a global equilibrium. Such a world is wonderfully forgiving: no matter what knocks we deliver, the ball will always return to the bottom of the basin. The managing institutions can therefore have a *laissez-faire* attitude.

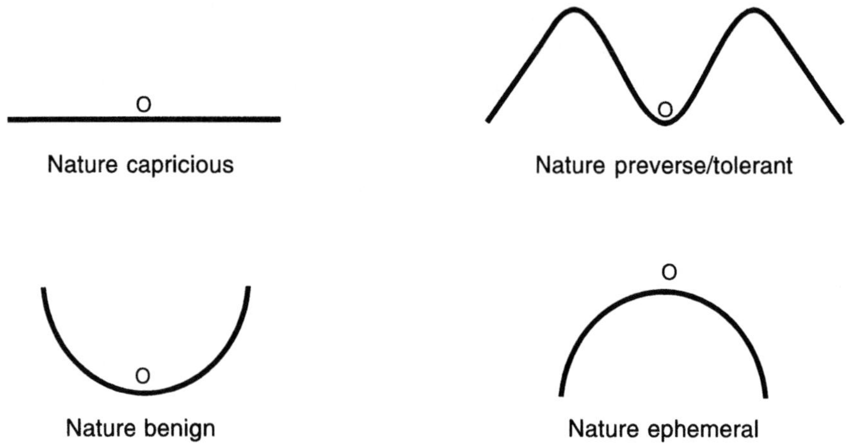

Figure.4.1 The myths of nature
Schwarz and Thompson (1990, p.5)

- *Nature ephemeral* is almost the exact opposite. The world, it tells us, is a terribly unforgiving place and the least jolt may cause its catastrophic collapse. The managing institutions must treat the ecosystem with great care. This is the view of the German Greens.
- *Nature perverse/tolerant*, though it may look like a cross between the first two, is quite different. Its world is one that is forgiving of most events, but is vulnerable to an occasional knocking of the ball over the rim. The managing institutions must therefore regulate against unusual occurrences. Our multinational is to be found here. It accepts that the small risk of disaster necessitates government regulation, but believes that, once minimum standards have been met, it should be free to make its own decisions.
- Finally, *nature capricious* is a random world; institutions with this view of nature do not really manage, nor do they learn. They just cope with erratic events' (Schwarz and Thompson, 1990, pp.4-5).

The authors add a typology of perspectives to the scheme presented above. The four myths of nature are related to the following four social types: 'Nature benign' is the perspective of the *individualist* who believes that problems can be solved by the regulating effect of the market-mechanism, whereas 'Nature ephemeral' is the conviction of the *egalitarian*, who believes that small is beautiful and who participates in small-scale organizations. This contrast might be related to our earlier presentation of ideologies. The 'nature benign' perspective fits within the dominant liberal market economy of the rich developed countries. 'Nature ephemeral' belongs to the tradition of anarchism. The other two social types, the *hierarchist* who believes in 'nature perverse/tolerant' and thinks that strong social controls are necessary and the *fatalist* who supports the idea of 'nature capricious' and is cynical about any political influence on social

processes are less easily to fit into our ideological scheme. The hierarchist, as described by the authors, can be associated with both conservatism and social democracy. The fatalist is too cynical to adhere to any political ideology.
We have presented in outline the way in which Schwarz and Thompson relate different ideas about environmental risk to social ideologies. Many seemingly scientific controversies about the risks of nuclear energy, genetic engineering and other effects of human-induced environmental change are based on much deeper and more fundamental differences in world-view. Lack of understanding of these different world-views can lead to fruitless 'scientific debates' and consensus-building on environmental problem solving.(see also section 7.8).

4.9 Continuing controversies

We have stressed the variety in opinions about 'good nature' and 'bad environment' time and again. Yet it makes sense to speak about 'environmentalism' as a widely shared and permanent criticism of basic trends in modern, western societies. 'Environmentalism' has become a legitimate form of opposition against the actual development of liberal market economies. We will describe this opposition by discussing four controversies.

4.9.1 Ecocentrism versus anthropocentrism

We started this chapter with a discussion about the way in which Christianity placed man above all other forms of life on earth. Modern evolutionary biology gives the human species a much more modest place in nature. However, the dominant way of thinking about life on earth can be called extremely 'anthropocentric'. This qualification even applies to trend-setting documents of environmental concern. Both 'Limits to growth' (Meadows *et al.*, 1972) and 'Our common future' (WCED, 1987) are only interested in the future perspectives of human life on earth. 'Environmentalism' as an ideology of concern about man-made environmental change criticizes this one-sidedness and adheres to an 'ecocentric' view of life on earth, embracing all forms of life instead of merely human beings. It goes far beyond the scope of this book to discuss the question of how far protection of non-human life on earth could and should go. Should every single plant or animal be protected or only the survival of species be guarantied? Notwithstanding these uncertainties the criticism of the one-sided interest in human well-being is a recurrent issue with a long history and probably a long future.

4.9.2 Steering versus engineering

We discussed above two contradictory views on human influence on environmental quality. The myth of nature called 'nature benign' justifies individual freedom in influencing and transforming nature, whereas 'nature ephemeral' requires detachment and care in human activities which have environmental effects. This controversy is relevant to the understanding of controversies about the development of agriculture and energy supply, to mention just two major sectors of society. Modern agriculture, with its high input of artificial fertilizers and pesticides and its transformation of species by means of genetic engineering, is severely criticized by environmentalists. Instead of this environmentally damaging type of agriculture, 'ecological' or 'biological' agriculture is advanced. This type of agriculture uses less environmentally damaging inputs and replaces them by manpower and biological measures like the introduction of plant and animal species which can prevent damage to crops by insects. In the field of energy supply, environmentalists try to reduce or even ban both the use of fossil and fissile fuels by a combination of energy conservation and use of renewable energy sources like hydroelectric power, sun, wind and biomass. The common element of these alternative methods of agriculture and energy supply is the preference of 'steering' above 'engineering'. Instead of modifying and exploiting nature, nature is steered to fulfil human needs. This difference in opinion about the way nature should be handled in order to fulfil human needs, strongly influences controversies about the development of agriculture, energy supply and other human activities.

4.9.3 Limits versus growth

In section 4.2.2. we discussed the contrast between the belief in progress and the cyclical notion of time. These differences in world view have far-reaching implications for the meaning of life and death.

Max Weber on 'progress' and the meaninglessness of death.

In his famous speech 'Science as a vocation' the German sociologist Max Weber argued that the western belief in progress degrades death into a meaningless event: Abraham or any other peasant of ancient times died 'old and satiated with life', because he stood in the organic cycle of life, because his life by its very nature had given him at the end of his days what it had to offer, because he had no more puzzles which he wished to solve and therefore he could have enough of life. But a civilized man, who is put in the midst of the continuing enrichment of civilization with thoughts, knowledge and problems, can become 'tired of life', but not 'satiated with life'. He snatches only the tiniest part of what the life of the spirit constantly produces, and then only something provisional, rather than final; thus death is for him a meaningless occurrence.

Weber, 1989, p.14.

In chapter 2 we discussed the studies of Meadows *et al.* and Hirsch on the ecological and the social limits of growth. The ambition of high and continuous economic growth has been criticized by environmentalists time and again. In reaction to that criticism, it has been suggested that 'de-materialization' of economic growth enables economic growth and reduction of the use of natural resources and emissions of polluting matters to go hand in hand. Notwithstanding the possibility of creating such 'win-win' situations, there remains a contradiction between the ambition for growth and the acceptance of limits. Environmentalism can be considered as a manifestation of a world view in which improvement of the quality of life is sought within given boundaries instead of by removing boundaries. Examples of such an 'accepting' attitude are the insight that death should not be postponed as long as possible under all circumstances, that travelling takes time and that the possibility of reaching every place on earth within a few hours leads to a great loss in the pleasure of travelling, that the on-going replacement of manual and mental activities of men by technical devices generates unemployment and dependence on equipment which are far beyond the understanding and control of their users. 'Environmentalism' does not ask for a 'frozen technology' and a 'frozen society' but replaces the strife for quantitative growth by an ideal of qualitative improvements within ecological, economic and social limits

4.9.4 Free nature versus prized products

Free availability of natural resources has often been mentioned as a major source of environmental problems (see section 5.2) and financial instruments (see section 9.3) are considered as an indispensable element of environmental policy. Yet environmentalism can also be interpreted as a revolt against the over-valuation of prized goods and services and the undervaluation of goods and services without prizes. Again more is at stake than merely the protection of the *physical* environment. It can be argued that liberal market economies suffer from an underestimation of both the meaning of free available nature in the form of clean air and water, open space, ecosystems, animals and plants and informal forms of help and social control, love, friendship and ceremonies. 'The environment' is an issue that legitimates protests against a neglect of those things in life which people often cherish most but which are absent in the news about Wall Street or the parliamentary budget.

Chapter 5
Social dilemmas

5.1 The essence of environmental problems: transfer of disadvantages

Environmental issues are socially determined. By definition, human intervention in the physical environment lies at the heart of such problems. Why do activities continue largely unchecked, even after it has become clear that they are harmful to the environment? There are apparently forces at work in our society which ensure that activities which are extremely damaging to that society nevertheless continue to take place. Environmental problems arise because the detrimental effects of a particular intervention invariably involve a different section of society to the one which benefits from that intervention. In other words, there are always groups in society which have a vested interest in precisely those activities which are harmful to others. Vested interests play a decisive role in the origin of environmental problems. An interest may be defined as something which is to the benefit of an individual or a group, and is therefore worth pursuing. The pattern we systematically recognize in the nature of environmental problems is an unequal distribution of advantages and disadvantages linked to the activities causing these problems. This is reinforced by the fact that the groups which suffer as a result of damage to the environment are not in a position to prevent others from causing that damage. The separation of advantages (in the form of the benefits resulting from human intervention) and disadvantages (in the form of environmental damage) is the ultimate cause of all environmental problems.

This separation of advantages and disadvantages means that the costs of environmental damage are passed on. While the benefits accrue to the one that performs the activity, the costs are transferred to the community, or to other parts of society that are not in a position to prevent this from happening. An example in terms of this 'transfer problem' can be found in the transitions in the agricultural sector. When we try to analyse the possibilities for a more sustainable form of agriculture, we will recognize many opposing interests. Recent developments in agriculture included enormous increases in scale, high intensity cattle breeding and arable farming, high investments, artificial fertilizers, imported cattle feed etcetera.

An example of the transfer problem.

In the agricultural sector the advantages of new developments have for the most part been transferred to:
- consumers (lower prices);
- suppliers and processors;
- developers;
- banks (increased foreign capital);
- the civil service (planning departments);
- prosperous farmers;
- and feed manufacturers (often in developing countries).

The costs are transferred to:
- those farmers who have ceased farming, as well as a great many of those still active;
- the collective sector (workers laid off);
- agriculture (reduction in soil fertility, increased vulnerability, reduction in biodiversity);
- water supply sector;
- recreation sector (stench, pollution of surface water);
- people involved in forestry;
- people valuing the countryside;
- animals (welfare of livestock).

Source: WRR (1984; Dutch Scientific Council for Government Policy).

Thus while the agricultural sector has increased its efficiency, which is in the general interest, there has also been an extensive transfer operation, with clear winners and losers both within and outside farming. In short, society passes the disadvantages on to the farmer, who in turn passes them on to livestock, the countryside, the environment, other sectors, other farmers and, not least, to the farmers of tomorrow.

The form which this transfer of disadvantages takes may vary considerably. In the case of environmental problems, it can take three forms: geographical, over time, and from the individual to the collective.

5.1.1 Geographical separation

The separation of advantages and disadvantages is often spatial. Much of the pollution of air, rivers and seas causes problems which go beyond the boundaries of a single nation, a problem that has been discussed in chapter 3. Thus a large portion of the air pollution originating in the Netherlands ends up in Germany and Scandinavia (acidification of forests and lakes). Similarly, as the Netherlands lies downstream, the chemical pollutants in the Rhine, the Meuse and the Scheldt end up in the Netherlands and the North Sea. The costs related to the discharges from the French potash salt mines are transferred to downstream water purification plants and consumers of drinking water. The possibility of passing on the harmful effects of pollution by means of a nearby river is often an important consideration in the choice of a location for potentially polluting installations. When nuclear power plants are planned, there is a noticeable preference for locations near the border; not only does the river provide cooling-water, but waste water flows out of the country. In general rivers that do not leave a country are cleaner than those crossing a border. France for example, is a country that has an extensive program for cleaning water, but the program concentrates on rivers with a home river basin.

Spatial or geographical separation is not only an international problem; similar processes also operate on a considerably smaller scale. The processing of household refuse, whether by dumping or incineration, is usually concentrated in a particular locality. The local communities bear the brunt of the disadvantages, while the benefits, in the form of the cheap and simple disposal of refuse, go to households elsewhere in the country.

5.1.2 Separation over time

There are a great many environmental problems which involve passing disadvantages on to future generations. There is a separation between advantages and disadvantages in chronological time: *Après nous le deluge* (after us the flood). Many activities of society are directed towards fulfilling short-term needs, while many forms of environmental damage do not manifest themselves until much later. The concept of sustainability (section 3.6) is based primarily on the time dimension of transfer of social costs to future generations. A problem which neatly illustrates the time factor in the separation of advantages and disadvantages is the depletion of raw materials. If future generations find themselves in a new Iron Age, they will be paying the price for the depletion of the world's resources which is presently taking place.

Waste is another environmental problem in which time plays a crucial role. We know that there are products and production methods which in the long term may be expected to cause serious problems. In the case of products, it is generally accepted that the advantages (at the consumption stage) outweigh the disadvantages (at the waste stage). The problems created by the waste stage are traditionally 'postponed', and we ourselves are faced with the waste of the past, in the form of such things as illegal dumping grounds for poisonous substances. If the waste problems which we pass on to future generations are to be kept within bounds, then we must look first to the design and manufacturing stage, keeping in mind that the product involved will eventually become waste.

A similar situation presents itself in the case of radioactive waste produced by nuclear power plants. Although the general concensus is that no satisfactory solution has yet been found for the problem of storing nuclear waste, in many countries this has not led to the conclusion that for that reason it would be better not to make use of nuclear energy for the generation of electricity. In most countries using nuclear energy provisions have been made for temporary storage, and no decision has yet been made on future measures. The present cheap production of electricity is accompanied by certain risks and costs which future generations will have to deal with. The low cost of that electricity has everything to do with the fact that other costs such as those involved in the conservation or dismantling of nuclear power plants will not become apparent for many years. In many cases these costs will be quite considerable indeed (Mackerron, 1989). Another case in point is the diffusion of non-degradable pollutants. The addition of phosphates to fertilizer, and copper to pig feeds, means that these substances enter the soil in a non-concentrated form and cannot be reclaimed. If continued, this practice is expected to cause enormous problems for future generations, particularly in the Netherlands, which has an enormous stock of pigs and therefore a large manure problem.

5.1.3 Individual advantages versus collective disadvantages

In the case of spatial transfer and separation in time, disadvantages are passed on fully. The results of many other environmental problems, however, are also borne by those responsible. In these cases the individual advantages generally outweigh the individual costs, mainly because the benefits accrue fully to the individual while a collective defrays the costs so the individual only pays a small part of it. Almost all forms of pollution which are concentrated geographically fall into this category. This is true of the chemical pollution of the environment referred to above, in spite of the fact that the consequences do not become apparent until later, and there is in effect a separation over time as well. There are other environmental effects which make themselves felt directly, such as air and noise pollution. Anyone responsible for the emission of harmful gases through the chimney or in the form of exhaust fumes is polluting everyone's environment. The contribution of a single motorist to air pollution is relatively small, as is the noise pollution produced by a single automobile. But the benefits provided by that automobile are reserved exclusively for the motorist himself. In all such cases, where the detrimental effects on the environment are more or less immediate, the burdens to the 'perpetrator' are always smaller than the advantages he enjoys, precisely because the burdens are passed on to society as a whole.

5.2 Costs which are not reflected in prices

There are a great many environmental costs connected with the production, consumption and disposal of various goods which are not incorporated into the price, and thus are not borne by the producer or the consumer. This phenomenon is essential in the analysis of the costs and benefits of polluting activities. The separation of costs and benefits creates the possibility of a 'transfer' of costs. These costs will ultimately be paid, but it is not clear when or by whom, but the fact that they will probably be paid by others rather than the one who has the benefits, there is strong tendency to ignore them.

5.2.1 External costs

Since Pigou (1920) described the phenomenon of costs that are not reflected in prices, these have been variously referred to as 'social costs' or 'hidden costs'. In economics the loss of welfare for the victims of a transfer of disadvantages is referred to as an external diseconomy or a negative *externality*.
As a result of the increasing gravity of our environmental problems and the growing interest expressed in all quarters, the term 'external costs to society' has become part of the language of economics. An example of these costs can be found in the work of Hohmeyer (1992) on the costs to society involved in electricity consumption. The various methods for generating electricity were compared, with and without the costs to society. The results indicate the extent of the transfer of costs to other sectors, in particular the society of the future. The benefits of less polluting ways of electricity generation become apparent when the social costs can be avoided by replacing fossil fuels and nuclear energy by renewable sources, such as wind and photovoltaic solar energy.

External costs

When the quantified costs of conventional energy systems for the production of electricity based on fossil fuels are totalled and standardized for the production of 1 kWh, gross social costs in the range of 0.03 to 0.16 DM/kWh result (1989DM). For electricity generated in nuclear reactors (not considering fast breeder reactors) gross social costs in the range of 0.1 to 0.7 DM/kWh result. (..) When one considers the social costs and benefits of electricity generated by wind energy - with the social costs of present electricity generation included as avoided costs - total social benefits in the range of 0.05 to 0.28 DM/kWh result.(..) The sum of net social benefits for photovoltaic electricity supplied to the public grid lies between 0.006 and 0.35 DM/kWh. (..) All assumptions underlying these figures minimize the advantages of renewable energy sources *(so they are conservative estimates; et/mw).*

Hohmeyer, 1992, p.367-368.

Actual electricity prices about 0.20 DM/kWh (=US $ 0.15).

5.2.2 Uncertainty about the costs to society

The energy sector is of crucial importance when it comes to the costs to society.

Another sector which plays an important role in industrial society is transport, which also involves enormous costs to society. A study carried out by Grupp (1986) in West Germany showed that these may be as high as 20 to 26 thousand million Euro (±30 to 40 US $), or about 350 Euro (±500 $) per head of the population per year. Why have no measures yet been taken to incorporate these social costs into the price, despite the magnitude of them?

Unfortunately, in almost all sectors there is a great deal of uncertainty with respect to the actual costs to society. This is reflected in the wide margins used for the estimates in the external costs, as we can see in the results from Hohmeyers' calculations (p.87). We know quite a bit about the advantages, because more research has been done into this aspect, and because they are manifested in the short term. The disadvantages, on the other hand, are diffuse; they do not become apparent until much later, and in certain cases, they are even unknown. Although the data of Hohmeyer do in any case show that the price now being paid for electricity does not even approach the ultimate cost to society, all the methods in use to estimate these costs are open to discussion. Beside the methods, there is always a discussion about which aspects should be taken into account. In particular about the social costs of activities that are not yet widespread, such as the application of renewable sources. For hydroelectric power we now know that the costs to society are in some cases much higher than expected in the past. With regard to the social costs of wind turbines Hohmeyer only quantified 'noise' as environmental impact. Probably this is also an underestimate, as in the public valuation the impact on landscape and scenic beauty is the most important factor (Wolsink, 1989, p.203). However, in an attempt to quantify external costs of wind power Eyre concludes that for specified areas the costs of visual impact are insignificant, although 'in areas of national scenic importance and high recreational use, visual amenity external costs could be large.' (Eyre, 1994, p.80). This shows the dilemma of quantifying external costs: these are highly dependant on the conditions of the application. Often external costs are assumed to be insignificant, because in specified conditions they actually will be insignificant. Nevertheless, in many cases the conditions do not fit these 'specified conditions'. In the case of wind turbines, they are sometimes actually sited in highly sensitive areas, as the Spanish example of bird collisions shows for yet another environmental impact of wind turbines that is often assumed to be insignificant if the sites are carefully chosen.

As a result of all the uncertainties about the extent and nature of environmental costs, there is often a tendency to disregard them during decision-making. Moreover, not all costs can be expressed in terms of money. However, it would not be true to say that environmental pollution is due exclusively to ignorance of the effects. The driving forces behind environmental pollution must be sought in the conflicting interests which arise out of the fact that it is possible to pass on to others the cost of environmental damage.

Wind farms blighting the landscape

Britain's big increase in wind turbines is proving a blight on the landscape, a report by a countryside organisation claims. Developers are exaggerating the output of the new 'green' power stations which are being allowed to proliferate with little control, says the Campaign for the Protection of Rural Wales. It calls for much tougher scrutiny by local councils before turbine schemes are granted planning permission, saying that despite claims by wind power companies, stations are producing only about 25 per cent if their potential. A huge number will be needed to make any real contribution towards reducing atmospheric pollution.

The Independent, April 18, 1997.

5.3 The environment as a collective good

The conservation of the environment seldom benefits only the individual; it is almost always to the good of the entire society. Environmental components all serve one or more functions in society. The sociological concept of function stands for an objectively observable consequence or effect of human activity. An environmental function is an effect or result of circumstances in the environment which is of significance for society. In most cases this means the practical value of certain environmental components, such as clean air to breathe, natural resources to exploit, or simply scenic beauty to enjoy. In an economic sense each of the various functions of the environment for society can be considered a commodity or a 'good' for which there is demand and supply. Although there may be considerable differences between these environmental goods, they are almost always public or *collective goods.*
The supplies of natural resources are environmental goods, thanks to the function in society they fulfil. If we consider the environment as a collection of goods, the question then arises how such goods are produced and consumed. In modern western societies we are capable of producing highly sophisticated products and distributing these products effectively. We manage to fulfil the most varying needs and wishes of consumers, in many cases at remarkable low costs. Now the question arises, why we are less capable of managing environmental goods?

5.3.1 The nature of collective goods

In the case of many environmental goods, the term production is somewhat misleading. We are talking about the social function of parts of our natural surroundings, goods that are already present. Nature has provided us with clean air, thanks to natural processes such as the production of oxygen by means of photosynthesis. From an economic point of view, the production of clean air consists of conserving forests, especially tropical rain forests, and limiting the extent of pollution. In many cases there is no question of 'production', because the environmental components are not renewable. Once it has been used, natural gas has gone; moreover, the supply is finite. The only solution lies in the economical use of such natural resources. Often the 'production' of environmental goods consists simply in restricting environmentally damaging interventions, or reducing the effects of such interventions. In the case of renewable environmental goods it is important to see to it that the regenerative capacity is not affected. In general, therefore, it would be more exact to speak of the 'conservation' rather than the 'production' of environmental goods.

Whether goods are collective or not is determined by the demand side. Collective goods are those goods that, when consumed by one person in a group, cannot be denied to others in that group (Olson, 1971, p.14). Many functions of the environment are such that the possibilities for consumption are unrestricted, or in practice difficult to restrict, for example in the case of air. Other environmental goods have certain restrictions, but cannot be totally denied to individuals (space). Often it is theoretically possible to restrict use, but the right to make use of the good in question is anchored in history (drinking water). In general we can say that while traditionally all environmental goods are collective, through the ages the rights attached to a number of these goods have been restricted (also see section 3.3). Thus there are restrictions on most forms of hunting, while the use of natural resources is usually subject to certain restrictions as well.

Environmental damage reduces the availability of environmental functions, and thus the availability of collective goods. Such collective goods must be conserved, and this requires specific contributions, generally in the form of restricted consumption or measures to ensure proper management. The crux of the problem is how individual consumers can be motivated to make such contributions, when the goods cannot be denied them precisely because they are collective goods. Economic analyses have led to the conclusion that the conservation of collective goods must be realized by means of government measures. This is something which cannot be left to the 'invisible hand' of the free market. In the following sections we will examine in more detail the nature of environmental problems, with special reference to the problems which arise as a result of the functioning of the free market.

5.3.2 Economic valuation

In economic analysis the transfer of costs is considered to cause a general loss of welfare. By definition an external cost only exists when an activity by an agent causes an uncompensated loss of welfare to others (Pearce and Turner, 1990, p.61). Theoretically there are two solutions to limit the loss of welfare. According to the theory of Coase (1960), an optimum of externalities can be reached when the victims of pollution own the right to use a natural resource, including the right that this resource is not damaged by any polluter. If a market is created for these property rights, whether they are held by the community or by private individuals, they can be sold. Then the polluter and the one who is suffering from pollution can bargain over the externalities. The right not to be polluted can be sold to a polluter at an optimum market price. However, this theory is only valid in perfect market conditions. Because of several 'market failures' such as limited competition, large transaction costs and practical problems in identifying the bargaining parties, in practice this optimum will often not be reached (Pearce and Turner, 1990, pp.73-77). In particular when the ownership of property rights is common and when there is no geographical separation involved, nor separation of costs and benefits over time, there is a problem that the agents selling property rights and the agents buying them are the same parties. In practice the market is not able to create conditions that limit environmental damage and therefore other instruments are needed. A big problem is that the value of environmental functions is unknown. There is common sense about the value of clean air to breath. Nevertheless, clean air is not for sale, it is simply there when it is not polluted. It is not produced, it is a product of nature which should be maintained and fostered. What is the value of refraining from polluting activities? Economists try to describe this value in terms of money, to make it comparable to other goods. The value of products is reflected by their price on the market, as anyone who is buying the product for that price is showing his *'willingness to pay'* (WTP). If there is no market where we can buy the 'refraining from polluting activities', then the value of it cannot be reflected in a price. However, this does not mean that people have no willingness to pay for it, but the political process should substitute the private market as the system of the provision of these collective goods (Seneca and Taussig, 1979 p.100).

In practice the valuation of environmental functions is the result of the political process, but it remains uncertain how these results of collective decision making correspond with people's actual individual willingness to pay. Methods for estimation of this valuation are being developed, such as *contingent valuation*, a method based on public surveys (Mitchell and Carson, 1986). These methods are still under discussion, however, and estimates show large differences. An example is the valuation of some kind of endangered species. An estimation of willingness to pay for the prevention of the loss of it will be very different when the assumption is that only people to whom the existence of that particular animal is important should pay for the prevention, compared to an estimation based on the assumption that all people in their role of tax-payers should pay for it. Yet another option is to look at the willingness to pay by people who are responsible for the loss (Harris and Brown, 1992). The conclusion is that methods like contingent valuation cannot offer undisputable indications for pricing environmental goods and the valuation remains in practice the result of collective decision making.

5.4 Social dilemmas

It is a commonly accepted view that the conservation of environmental functions requires vigorous government measures. Environmental problems are rooted in processes of transfer, and the intervention of those in authority will be needed to counter those processes.

5.4.1 The 'tragedy of the commons'

In one of the classic texts in environmental literature, this view is illustrated by means of a simple example. It was put forward by Garrett Hardin (1968) primarily to demonstrate the need for a population policy, but it is equally effective in illustrating the way we deal with environmental goods.

Tragedy of freedom in a Commons.

The rebuttal to the invisible hand in population control is to be found in a scenario first sketched in a little known pamphlet in 1833 by a mathematical amateur named William Foster Lloyd (1794-1883). We may well call it 'the tragedy of the commons,' using the word 'tragedy' as the philosopher Whitehead used it: 'The essence of dramatic tragedy is not unhappiness. It resides in the solemnity of the remorseless working of things.' He then goes on to say, 'This inevitableness of destiny can only be illustrated in terms of human life by incidents which in fact involve unhappiness. For it is only by them that the futility of escape can be made evident in the drama.'
The tragedy of the commons develops in this way. Picture a pasture open to all. It is to be expected that each herdsman will try to keep as many cattle as possible on the commons. Such an arrangement may work reasonably satisfactory for centuries because tribal wars, poaching, and disease keep the numbers of both man and beast well below the carrying capacity of the land. Finally, however, comes the day when the long desired goal of social stability becomes a reality. At this point, the inherent logic of the commons remorselessly generates tragedy. As a rational being, each herdsman seeks to maximize his gain. Explicitly or implicitly, more or less consciously, he asks, 'What is the utility *to me* of adding one more animal to my herd?' This utility has one negative and one positive component.
1) The positive component is a function of the increment of one animal. Since the herdsman receives all the proceeds from the sale of the additional animal, the positive utility is nearly +1.
2) The negative component is a function of the additional overgrazing created by one more animal. Since, however, the effects of overgrazing are shared by all herdsman, the negative utility for any particular decision-making herdsman is only a fraction of -1.
Adding together the component partial utilities, the rational herdsman concludes that the only sensible course for him to pursue is to add another animal to his herd. And another; and another ... But this is the conclusion reached by each and every herdsman sharing a commons. Therein is the tragedy. Each man is locked into a system that compels him to increase his herd without limit – in a world that is limited. Ruin is the destination toward which all men rush, each pursuing his own best interest in a society that believes in the freedom of the commons. Freedom in a commons brings ruin to all.

Hardin (1968) p.1244. Reprinted with permission of Science © 1968, American Association for the Advancement of Science.

According to Hardin, the solution to environmental problems related to the use of collective goods which he attributes mainly to the growth of population ('freedom to breed', p.1246) lies in 'mutually agreed coercion' (p.1247). This not only includes constraints laid down by government, but also those formed by the privatization of collective goods. Following on from there, we could say that if we accept that there is such a thing as an exclusive private right to certain lands, for instance, then others can if necessary be denied access to those lands. Hardin's example has a certain appeal, as witnesses the role which his article has come to play in environmental literature. Nevertheless, the analysis as well as the accuracy of the description of the tragedy has triggered no small amount of discussion. As Cox (1985) points out, Hardin's version of the commons tragedy is historically inaccurate, in particular the idea that traditionally everyone had a right to the free and uninhibited use of the commons. This notion may correspond to our economic definition of collective goods. However, the situation on the commons was quite different at the time Hardin is referring to. The freedom to make use of common land was restricted to a few specific rights of use, and then only to the extent that these were essential for one's livelihood (Cox, 1985, p.54). The intensive use of land for one's own gain had no place in the concept of the commons. The idea that the use of common lands was unregulated is also contrary to fact. The commons have indeed been lost, but in a manner which runs counter to the description which Hardin has borrowed from a nineteenth century amateur mathematician, W.F. Lloyd. Hardin considers the enclosure of the common lands as a remedy for overgrazing. In sixteenth-century England the commons were appropriated by feudal barons and the other members of society were gradually deprived of all rights of use. The intention of the landed gentry in enclosing these lands was precisely what Hardin accuses the other users of: the economic exploitation of the lands (Hobsbawm, 1979, p.101). The lands were used to graze sheep on, producing wool for the Flemish wool manufacturers, and the rest of the community was deprived of the use of the lands. The famine was not due to overgrazing by farmers in general, but to sheeprearing by the landed aristocracy and the denial to others of the rights of use. The enclosure also came to play an important role in the industrial revolution, for the famine resulted in a mass drift to the cities, ultimately providing a supply of cheap labour for the new industries (Hobsbawm, 1979, p.104).

5.4.2 Hobbes' State of Nature

The fundamental problem of Man's interaction with his fellow human beings and with the environment is a dilemma put forward by the seventeenth-century philosopher Thomas Hobbes. He describes the natural circumstances and necessities of human beings. In Hobbes' primitive society there is no question of either culture or community. Everyone is concerned with his own interests, and in any relations with others, one's own survival is of primary importance.

According to Hobbes, human nature is such that the fate of others is not a matter of concern, so that in effect each individual forms an unknown but undeniable danger to his fellows. The consequences of this fact are far-reaching: the total lack of civilization. In the words of Hobbes:

> 'No commodious building, no instrument of moving or removing such things as require much force, no knowledge of the face of earth, no account of time, no letters, no society, and which is the worst of all, continuall feare and danger of violent death. And the life of man, solitary, poor, nasty, brutish and short.' (Hobbes, 1971, p.65).

What Hobbes was describing here was the paradoxical consequence of Man's instinct for self-preservation. A simple example will illustrate what he meant by a war of 'all against all'.

In a State of Nature two hunters meet. Both have two simple aims: to survive and to conquer. But survival is more important than conquest: better a slave than a dead man. The two of them meet during the hunt and both are armed. At first they do not know exactly what to do, for both survival and conquest will depend on what the other man does. Before launching an attack, they each try to work out what the various options are. Figure 5.1 shows all the possible results of such a meeting.

		Hunter B	
		Without arms	***armed***
Hunter A	***Without arms***	A and B make peace	A dead B Victory
	armed	A Victory B dead	A and B both hurt

Figure 5.1. Decision options for two hunters in a 'state of nature'.
Source: Schmidt and Van der Veen (1976).

All four options will be carefully considered by both A and B. It is clearly a question of opposing interests here, in that the conquest of one of them will mean the death of the other. Since they both have a preference for staying alive, they agree that the situation in which neither of them is armed is preferable to the situation in which both of them are armed. One of them proposes that they count to three and then throw away their weapons. The other agrees, but then they both start to wonder whether they should keep their word. From the stand-point of each of them separately, it is clear that they should not. One of the hunters reasons 'Whatever he decides to do, I'll be better off if I hang on to my weapon.' The same thing goes for the other man, and both decide not to throw down their weapons. As a result they fight it out, and the desired situation of peace is unattainable.

One may wonder whether this characterization of primitive Man is correct. Unlike Hardin, Hobbes did not pretend to describe a historically well-founded situation. He was interested not so much in the historical accuracy of his description as in the analysis (Sabine, 1968, p.464). In any case he believed in his analysis: apparently he believed that this is what life among primitive beings would be like. In that analysis he stated that the only way of escaping such a bleak prospect was by force, i.e., a mighty sovereign power capable of exacting the necessary cooperation. The fact that he was describing the natural characteristics and circumstances of all human beings makes his analysis relevant at all stages of civilization. With respect to their relationship with their own surroundings and those of others, human beings have not changed. If we are to escape the disastrous consequences of our interaction with the environment, then a 'civilizing' process will have to take place (see section 4.5). Many of the collective goods which are part of a civilization have indeed been obtained. But when it comes to preserving the environment, that process of civilization has not been carried far enough. This will not come about of itself: it will have to be enforced by someone in authority.

This hypothetical example bears a close resemblance to Hardin's 'tragedy of the commons', just as the solution put forward by Hardin also resembles that of Hobbes. It is only by means of force, constraints on freedom laid down from above, that human beings can be made to conduct themselves in a civilized fashion. Hobbes' theory of government is based on his vision of the true nature of Man. That government is rooted in a 'social contract', similar to the agreement on mutually agreed constraint described by Hardin.

Hobbes was describing a fictitious primitive human being, whose nature is such that it is immediately clear that we are not dealing here with a civilized society. Hardin's case is different: the users of the commons lapse from a relatively civilized situation into one in which they destroy their own collective good. Hardin attributes this to the increased claims on the common good made by individuals. He proceeds on the assumption that from an economic point of view, human beings are inclined toward selfishness. The only way to escape is a social contract, the basis of any kind of civilized society, regulating the obligations of any individual toward the requirements of all other individuals within community. Hardin assumes that when it comes to common goods, everyone has a right to help himself, without regard to the requirements of others. Even when certain needs have been met, one still has a right to more. Such reasoning contributes to the creation of shortages where these do not naturally exist (see section 2.5). There is a strong parallel between the example of the hunters in the State of Nature and the situation on the commons according to Hardin's description. The essential point is the paradox: if everyone acts on his own, in his own interest, everyone will ultimately lose out. Such situations are known as social dilemmas, a term which originated in the theory of games.

5.5 Game theory and society

Game theory is a decision theory, which formulates rules for optimization of outcomes in strategic choice situations. It centres on the behaviour of people or organizations (called players) in situations where decisions must be taken, where the conduct of one player influences that of the others, and where the result is determined by the conduct of all the players. Originally this was a mathematical theory about board games (Von Neumann, 1928), but since the publication of *Theory of Games and Economic Behaviour* by Von Neumann and Morgenstern (1947), it has increasingly been applied to social problems. The earliest of these applications had to do with economic decisions; there is little attention to suppositions about the motives underlying human conduct, and a strong emphasis on formalized mathematical reasoning. It was Luce and Raiffa (1957) who made the theory of games accessible to non-mathematicians. Psychologists and social scientists were the next to adopt the theory, although their attention was directed towards the assumptions, rather than the formal rules and outcomes. In other words, social scientists are more inclined than economists to ask themselves whether the assumptions about the preferences of the participants do indeed reflect reality. They ask themselves whether the true nature of Man is that sketched by Hobbes. As in the case of the farmers described in Hardin's example, we are dealing here with the supremely rational *homo-economicus*: only those actions undertaken in one's own interest are considered rational. Psychologist studying social dilemmas do not take this restricted interpretation of rationality for granted and they concentrate on variations in preferences and the existence of other values rather than economic rationality (chapter 6).

Hardin's article also helped to direct attention toward the theory of games in relation to the analysis of environmental problems. The effects of such problems usually affect individuals other than those who created them, because the effects have been transferred, in the same way as Hardin's farmers in the commons story transfer the costs of extra grazing to all users. The idea that many environmental problems manifest themselves because the earth and its natural resources are finite also helped to highlight the theory of games (Carraro and Filar, 1995). In many cases the intensive use of resources has resulted in a scarcity which did not previously exist, or was not recognized as such. It was primarily environmental problems which led economists to study the 'zero-sum society' (Thurow, 1980), in which any benefit to one person is invariably gained at the expense of others. Such a situation is characteristic of most games situations described by the theory of games: a win by one player is accompanied by a corresponding loss for the other players ('zero-sum games'). In a zero-sum society conflicting social interests are heightened. One party's gain is achieved by passing on the loss to others. Social dilemmas offer an excellent opportunity to analyse conflicting interests and problems of the transfer of costs.

5.6 Basic types of social dilemmas

The significant variables determining the structure of all social dilemmas, including the ones concerning environmental issues, are summarized in table 5.1. Assessing the nature of a social dilemma, the outcomes that will be most likely, and the possibilities for intervention, should always include all seven variables.

Table 5.1 Variables determining the structure of a social dilemma.

- ★ the set of positions (preferences) held by participants
- ★ the participants in each position (specified preference)
- ★ set of alternatives (decisional options) for participants
- ★ set of outcomes for all combinations of decisions
- ★ set of functions mapping actions of participants
- ★ the amount of information available
- ★ the costs and benefits assigned to actions and outcomes

Source: Ostrom, 1986.

However, in general there are a few basic categories of social dilemmas (Hamburger, 1979). To simplify matters, our discussions will be based on problems of choice with two alternatives: co-operation (C) and rejection (R). The classic example is he *Prisoner's Dilemma* (PD), which Luce and Raiffa (1957, p.94) describe as follows:

- ★ Each participant has a dominant strategy, i.e., a choice which under all circumstances results in a greater advantage to the individual than the alternative.
- ★ If all the participants choose the dominant strategy, then this will lead to a less satisfying result than would have been the case if all the participants had not chosen the dominant strategy.

One example of the PD is the situation described in figure 5.1, involving the two hunters. This illustration of the unhappy fate of Hobbes' primitive Man is a classic 2x2 PD game, a situation in which there are two players, who each have two options. This type of game, with two participants, is of mainly illustrative value; however, the PD can be extended to an n-person dilemma, also known as a *commons dilemma* (Dawes cs., 1977). The PD owes its popularity to the fact that it is deceptively simple. In all cases there is one dominant strategy, no matter what choices the opponents make. Even when there are more than two players, the same strategy is always dominant, no matter what choices the other players make. As a result, the game is often played in a highly simplified form, in which one individual plays against all the other players, who are together portrayed as a single player. Thus individual differences between the preferences of the participants are of no importance here. Dilemmas in which there is not in all cases a single dominant strategy do not lend themselves to this simplified form.

The *trust dilemma*, originally introduced as the assurance game (Sen, 1974) is an example of a dilemma in which there is not always a single dominant strategy. The choice of non-co-operation, is favourable for an individual provided the number of persons who decide to co-operate is below a certain level. Once a certain number of persons have opted for co-operation, it appears that the choice of co-operation is also favourable from the point of view of individual benefit. A person who wants to maximize his own individual benefits will opt for co-operation, provided he has sufficient confidence in the co-operation of the others. In this 'trust' dilemma the dominant strategy depends on the trust in the co-operation of others. A person who is himself faced with a choice cannot be certain of the co-operation of the others, and must therefore tale a gamble. If the level of co-operation is low, one is better off refusing to co-operate, but if a large number of players co-operate, then it is better to co-operate.

The phenomenon known as squirrelling is an example of the 'trust' dilemma, which is also applicable to various types of environmental problems. Young children who do not live in the direct vicinity of school are often taken to school by their parents, who feel that there is so much traffic that it would be dangerous to let them walk to school on their own. If the number of parents who take their children to school by car, instead of sending them on foot or by bike, is excessively large, this tends to make it more dangerous for the children who walk or cycle to school. Precisely because of all the cars driving to the front of the school to drop off or pick up children, more and more parents will decide to drive their children to school.

Another example is noise annoyance caused by neighbourhood disturbances. The collective goods 'peace and quiet' can only be attained if no one in the neighbourhood regularly makes an excessive amount of noise. In that case individuals are rewarded for restraining themselves. If one person decides not to abide by the regulations, then there is no peace and quiet, and it makes very little difference whether or not others are making an effort to keep down the noise level.

The term *chicken dilemma* is used for a situation where it is profitable for the individual not to co-operate, when a certain number of the other players have already opted to do so. This dilemma, which is very common, is nicely illustrated by commuter travel. Day in day out, commuters who drive to work subject themselves and others to irritation and delay as a result of tailbacks and traffic jams. While they prefer to travel by car, the tailbacks detract from the advantages of their favourite mode of travel. If they opt for co-operative behaviour by deciding to travel by train or bicycle, these motorists must give up their first preference. But this does relieve them of the irritation caused by driving in heavy traffic. If enough people decide to take the train, to cycle to work, or to start working on different times of the day, this will have certain consequences for the people who make no contribution whatsoever to the solution of the problem. If there is less traffic thanks to the co-operative behaviour of others, who have switched to another mode of travel, then the motorists who do not co-operate are also spared the daily irritation of tailbacks and traffic jams. They are now free to make use of their preferred means of travel: by car.

5.7 Environmental dilemmas

Collective goods may take a variety of different forms. These various forms determine the type of social dilemma which can be prevented by preserving such goods.

5.7.1 Inclusive versus exclusive goods

There are certain goods which can be consumed without restricting consumption by other people. But there are other goods which when consumed by one person, are less available to others. This is reflected in the distinction between inclusive and exclusive goods (Olson, 1971, p.38). A typical example of inclusive goods is a radio transmitter. Anyone can receive the programmes being broadcast without interfering with the reception of others. Many environmental functions are inclusive goods. When we enjoy breathing in clean, unpolluted air, for example, we in no way interfere with others' enjoyment of that same air. Parents who are able to build a sandbox for their children in the back yard without encountering the remains of a dumping ground for toxic waste, do not restrict their neighbours' use of that same non-toxic land. But there are a great many environmental functions -and these include the majority of natural resources which are exclusive, i.e., consumption by one person excludes consumption by others. Natural gas, for example, can be used only once to heat a house. To some extent, space is also exclusive, since the use by one individual limits the potential use by someone else. This will, however, depend on the type of use.

5.7.2 Continuous and 'lumpy' goods

A second distinction which governs the supply of collective goods is the difference between continuous goods and indivisible or 'lumpy goods'. Some clarification may be required here. There are environmental goods which can in principle be supplied in any desired quantity. Each section of forest that is planted (or not cut down) makes its own contribution to the purification of the air. Each tree takes up carbon dioxide and emits oxygen. Alongside such continuous goods, there are other collective goods which do not become available until a certain minimum contribution has been made on the part of society. The construction of a bridge is a good example here. The goods only become available when so many contributions have been made that the entire bridge can be built. Goods of this type, which are only of value if they are available in indivisible, mass units, are known as lumpy goods.

It is important to note here that many environmental goods are lumpy goods of a very special kind (Taylor and Ward, 1982, p.354). Our renewable natural resources, such as lakes, rivers, or even Hardin's commons, appear at first sight to be both inclusive and continuous. Use of the environmental functions is, however, only possible up to a certain highly critical level. Above that level, these goods become scarce, and thus exclusive. When consumption increases they are in danger of losing their value altogether, because their regenerative powers are affected. We are referring here to natural resources whose regenerative powers are affected by the destruction of ecosystems as a result of extensive or incorrect use, such as the discharge of pollutants. A lake provides a certain amount of fish for anglers, but if too many fish are caught, there will be nothing left for anyone. The stock of fish can also be affected by competitive use, such as the discharge of waste water.

The 'lumpy' nature of collective goods adds an extra problem to the social dilemma. In the case of continuous goods we can at least say that the contribution of each individual always results in an improvement. This is not true in the case of lumpy goods. The contribution of an individual is only beneficial if it is one which helps to reduce consumption to below the critical limit. The fish stock in the over-fished lake will not recover until the total number of fish taken out is reduced to below the critical level, something which the individual is incapable of doing on his own.

5.7.3 The origin of dilemmas

A certain form of behaviour which has been going on for many years without causing any particular problems may suddenly prove to be the non-co-operative alternative in a social dilemma. This is related to the fact that many environmental goods are lumpy goods. One formulation of the dilemma in which the manner of development is taken into account is the *social trap* (Platt, 1974).

In most cases the critical level is exceeded as a result of the enormous demands made on natural resources. Owing to increased economic growth, these demands have increased significantly. Motor traffic, winter sports such as skiing, and pig-breeding are examples of activities which, thanks to growing prosperity, have become so extensive that nature can no longer bear the burden. This signals the emergence of a social dilemma. However, a complicating factor here is the fact that behaviour that is now considered as detrimental to the environment is often a continuation of behaviour which in the past did not lead to problems. One of the factors contributing to such environmental dilemmas is social inequality. As the consumption of particular goods are limited, because only a small part of the population can afford them, an activity may remain without environmental consequences. As a result of the general increase in prosperity, the consumption of those goods may suddenly come within the reach of many more people, so that critical limits are exceeded. Someone who for years has watched other people driving around in cars will probably not see private car ownership as an environmental problem once he himself is finally in a position to buy a car.

The social trap also plays a role in a series of decisions which, when seen separately, are quite rational, but which taken together lead to socially unacceptable results. This phenomenon is known as 'entrapment' (Brockner and Rubin, 1985). It often occurs when large projects have to be financed (Teger, 1980). During the development stage of a project all sorts of 'hidden costs' come to the surface, such as those related to unforeseen environmental effects. In spite of continuing set-backs, the decision-makers resolve to soldier on, because so much has already been invested. The costs soar, ultimately reaching a figure which is so high that, if it could have been foreseen, it would have resulted in the project being rejected.

Nuclear plant to be leisure park

During the next five years the fast breeder nuclear reactor plant at Kalkar, a small town near the Dutch border and flanked by the river Rhine in the German state of North-Rhine Westphalia, will undergo an extraordinary metamorphosis. Buildings which once housed sophisticated scientific equipment will be turned into hotel rooms. Land once laced with electricity wires and pipes will become children's playgrounds. Storage sheds and administrative offices will be turned into sports facilities. (...)

Financing and construction costs amounted to DM8bn ($5.7bn): the state and federal government put up DM4bn of tax-payers' money; the Belgian and Dutch government as potential users each contributed DM1bn; the electricity companies DM1.6bn; and those producing the energy paid only DM400m. As construction proceeded the razzmatazz soon wore off. Public opinion started moving against nuclear power, especially after the Chernobyl accident in 1986, while the government of North-Rhine Westphalia became anti-nuclear as well.

The future of Kalkar was passed from local, to state, to federal level and back again. No one was prepared to take responsibility for its future. The plant, built at vast public expense, was never commissioned. (...)

Saddled with a redundant site of 300.000 square metres, the local authorities at Kalkar and RWE earlier this year advertised the sale of the plant in local and foreign newspapers. (...) An acceptable solution has now been found for Kalkar. After all, people have much more free time now.

Financial Times, November 8, 1995.

A striking example of entrapment is the fast breeder reactor in Kalkar, built jointly by (West-)Germany, the Netherlands and Belgium (see frame). Costs kept going up, owing to the increasing complexity of the construction; the risks had been underestimated, and again and again standards had to be adjusted. The Dutch contribution to the project was originally financed by a surcharge on all household utility bills known as the 'Kalkar levy'. The contributions of electricity consumers and taxpayers in the three countries continued to rise. The only reason that the reactor was ultimately completed was that otherwise all the money previously invested would have gone down the drain. The necessary permits were never issued, and in the end the whole project was dropped. The power plant in which more than US$ 5 bn had been invested was ultimately sold for a few millions to serve as Nuclear Water Wonderland ('Kernwasser Wunderland'), an amusement park.

5.7.4 The volunteer's dilemma

A special type of dilemma is that formed by the search for locations for community facilities such as landfills, waste incineration plants, coal-fired power plants, nuclear power plants, and so on. All such installations involve risks and costs which are concentrated locally, and this leads to a full-blown social dilemma (O'Hare, 1977). According to Thurow, this phenomenon leads to 'balkanization', whereby an area disintegrates into separate regions, all concerned exclusively with their own interests. Regions and local communities are less and less willing to bear the cost of projects designed primarily to benefit people in other parts of the country. This tendency is a characteristic of the zero-sum society (Thurow, 1980, p.14). Often the factors which cause such discontent are environment-related. People living near an incineration plant are afraid of toxic emissions, and in many cases such fears have proven to be well-founded. O'Hare *et al.* (1983) noted that in the United States no new plants for the disposal of hazardous chemical wastes had been built in ten years, despite the fact that there was a concensus on the necessity for such installations, in order to prevent illegal dumping. This phenomenon has been christened the not-in-my-back-yard (NIMBY) syndrome, and the installations are called locally-unwanted-land-use (LULU's). Because it is always a question of finding just one site, the common goods are guaranteed as soon as one location is prepared to have the facilities built on its land. This is known as the 'missing hero trap' (Platt, 1973) or the *volunteer's dilemma* (Diekmann, 1985). It is precisely because the collective aim is achieved when only one location has been found which will accept the installation, that this situation promotes non-cooperation. It is actually an extreme variation of the 'chicken' dilemma. In sections 7.6.2 and 9.8 the community responses and political problems of facility siting will be discussed.

5.8 Nuances in economic egoism

The solution to social dilemmas is often the result of quite pronounced assumptions about the preferences of those who are faced with such a dilemma. The most important assumption is concerned with the rational economic nature of these choices. What Hardin calls rational behaviour is in effect little more than economic egotism. In Hobbes' view such selfishness is simply part of human nature. Moreover, both Hardin and Hobbes seem to support the view that when it comes to common goods, everyone has a right to everything, without regard for others. But, as we have seen, the tragedy on the commons did not proceed in the manner described by Hardin. There is also something wrong with his proposition that 'everyone is caught within a system that forces him to go on increasing his herds without restraint'. It is not only the 'system', a basic assumption is that the actions are governed by pure economic egotism. In decision situations people do not always act rationally. This fact has been recognized by a number of economists, and new models have been developed for players who act in a semi-rational manner.

Rationality is a matter of preferences. In many applications of the games theory, especially the PD, the crucial assumption is that no one will give up his own interests. The economist Amartya Sen proposed the use of analyses which include not only preferences in line with one's own interests, but also moral preferences (Sen, 1973). Rationality as the basis of decisions in social dilemmas can also be contested on empirical grounds. One need only think of those situations in which individuals voluntarily pledge their co-operation, even though it is not rational to do so (Sugden, 1985). Given the same circumstances, people make different decisions. To the extent that their behaviour is rational, they clearly have different preferences. It is obvious that particular acts, often behaviour that should be classified as based on environmental values such as separating household waste, cannot be explained merely by the economic rationality. Another example, consumers voluntarily paying a higher electricity price, is given in frame (p.104). Summarizing, it is an oversimplification to assume that in the case of a social dilemma all the participants have precisely the same preferences. Their behaviour may be based on a wide variety of considerations.

The above criticism does not mean that game-theory has been undermined. It does, however, make the analyses more complex. When using game theory two things are important to notice. First it is necessary to avoid using the wrong models to analyze subtle social situations. The Prisoner's Dilemma game in most cases is certainly not a realistic model for the description of reality. Secondly, when we employ game-theoretic solution concepts the crucial points are always the assumptions about the knowledge, beliefs and preferences of players and about the beliefs about actions taken by other players. Considering the issue of new laws and regulations we should be aware of the fact that 'the more we have to make such assumptions, the less certain we can be that our model will accurately predict the way individuals behave.' (Baird *et al.*, 1995, p.13).

Consumers can soon sign up for green electricity

Britain's consumers will be offered the chance to buy green, non-polluting electricity from early next year. Several electricity supply companies are planning to launch schemes in which customers would sign contracts specifying that all their power came from renewable sources like the wind, water and methane gas generated by rotting garbage. These consumers would pay a premium price - perhaps as much as 10 per cent higher than ordinary bills.
The new thinking was unveiled at a major conference on global warming in London yesterday organised by the World Wide Fund for Nature (WWF). The fund is interested in auditing and endorsing such schemes, ensuring they actually boost renewable energy over polluting fossil fuel power stations. Electricity companies are interested in being able to use WWF's famous panda logo to help sell the scheme.

The Independent, June 5, 1997.

The situations described as pitting one player against all the others are only useful as illustrations. The opponents are all assumed to display the same behaviour, which is simply not a realistic assumption. The nuances of the rationality hypothesis make it even more important, when considering an environmental problem, to examine the exact nature of the collective goods, and the preferences of the people involved. The method whereby a single player is pitted against a uniform group of opponents leads to an over-emphasis on PD analyses and solutions (Axelrod, 1984). Many environmental dilemmas are not prisoners' dilemmas, but rather problems in which the proportion of people who are inclined to be co-operative as against those who are not, play a crucial role.
In the case of environmental problems, people are often quite prepared to co-operate, and in many cases even prefer to do so. But even then there is a problem, because large-scale co-operation is then only assured when people are certain that the others will also co-operate. In effect it is the 'trust' dilemma which is the crux of most environmental problems. Even when people have a preference for co-operation, the dilemma inherent in the situation often means that not enough people put that preference into operation. Let us turn again to Hardin's tragedy: the dilemma of the farmers did not arise until other people who made use of the common lands, i.e. the landlords, violated the centuries-old trust by not keeping to the rules. They started to take more themselves while excluding others.

5.9 Individual differences

Under the same circumstances, different people may make different decisions. This may be attributed to differences in character and personality. The crucial assumption in the commons dilemma is that all the farmers will allow their own individual interests to take precedence. Economic rationalism closely resembles simple egotism. But there may be other individuals, who are more altruistic and who may be expected to attach less importance to maximum fulfilment of their own personal aims. They will be more likely to opt for the common interest, or the interests of future generations. Moreover, in a sense these choices, too, are rational. It is important not to confuse interests such as self-interest, group interests and the interests of others with traits of character, such as egotism and altruism.

There are other types of personality which may also influence the behaviour of individuals faced with a dilemma. People may have different value orientations, resulting in for example altruistic, co-operative, individualistic or competitive motives for their decisions. In examining the transfer of costs, which plays such an important role in environmental problems, the personality and attitude of the individuals involved must be taken into account. Thus we may expect the average entrepreneur to be considerably more competitive than the average consumer. Environmental problems caused by dilemmas involving choices by, say, farmers, are not the same as those involving the choices of consumers at the local supermarket. This may explain the fact that it is much easier to obtain the co-operation of households in addressing the problem of household waste rubbish than it is to obtain co-operation of pig stockholders in solving the problem of the manure surplus.

5.10 Theoretical solutions

The 'solution' put forward by Hobbes and Hardin lies in a powerful and even in some respects authoritarian state. It is indeed tempting to see government as the body responsible for solving social dilemmas, although one may well ask whether the solution of environmental problems is reconcilable with democracy. According to the analysis of collective goods and co-operative action (Olson, 1971), the conditions for voluntary co-operation in social dilemma situations are bad. However, more recent analysis of the developments of collective or cooperative behavioral patterns (institutions) modifies that point of view (Ostrom, 1990). In chapter 9 we will be looking more closely at the question of whether the solution of environmental problems should be left to the authorities.

If the social contract is seen as the basis of the authority which is capable of enforcing co-operative behaviour, then the 'powerful state' solution may be seen as a simple and extreme interpretation of 'mutually agreed coercion' (Hardin).

However, acceptance of government measures is not always automatic. Do the authorities always exercise proper care in the conservation of our collective goods? Do they make use of all available means to safeguard those goods? Does power lead to a distribution of financial burdens which is felt to be fair? Is it really logical to see in the state the solution for our environmental problems? We will discuss these questions later in chapter 9.

Behaviour which is imposed is often doomed to failure. In many cases it would be better to make use of more subtle means, which fall short of the qualification 'enforced' behaviour. The most obvious of these are measures affecting the advantages and disadvantages to the individual. Thus financial measures can make non-co-operative behaviour much less attractive. Such economic measures are known as market-oriented measures. However, solutions of this kind involve more than simply altering the price of products or services. The degree to which people actually opt for co-operative behaviour will also depend on the manner in which the incentives are presented: as a reward for co-operative behaviour or as a sanction for non-co-operation. People tend to attach more importance to disadvantages than to advantages (Kahneman and Tversky, 1979). The choice of an automobile with a catalyst is an example of behaviour which can be influenced by the price mechanism. But the difference in price is more effective when it is presented as a surcharge for an automobile without a catalyst than as a subsidy for one with a catalyst. However, there is a catch. When faced with a dilemma, not everyone will display the preferences of what Sen calls the 'rational fool'. By introducing measures which stress the economically rational options, the authorities may alienate some people who were originally inclined to co-operate. Giving blood is a case in point: many people freely donate their blood, and the offer of financial recompense might be a deterrent rather than an incentive (Titmuss, 1970). For further discussion on this subject see section 6.7.1.

A third way of influencing behaviour in the face of social dilemmas is communication. Highlighting the consequences of certain behaviour is often quite effective. Energy consumption is one of the first types of behaviour with environmental consequences which was studied experimentally (Stern, 1976). A series of studies have demonstrated the effectiveness of general information on energy consumption and the possibilities for savings; feedback in the form of regular (weekly, monthly) reporting on the results of behaviour; and programmes to encourage consumers to keep a record of their behaviour (Samuelson, 1990).

In some types of dilemmas it has proved to be useful to improve the communication between those individuals confronted with the dilemma, and to increase their involvement in the problems which threaten to arise as a result of the dilemma. This provides additional insight into the reasons which lead others to opt for co-operative behaviour (Neidert and Linder, 1990). Such a strategy may be particularly effective in situations where a 'trust' dilemma is involved. The potential success of measures directed at improved communication will depend to some extent on the size of the group. The greater the number of people involved, the more difficult it is to solve the dilemma by means of communication.

Communication between individuals may have the effect of co-ordinated behaviour, which greatly increases the chances of getting others to co-operate, thus solving the fundamental problem of the 'trust' dilemma. The provision of most environmental goods are characterized by De Jasay (1989, p.57) as public issues that are mainly co-ordination problems. Although these are among the hardest problems society has to face, co-ordination is the key issue, not the tendency toward free riding and other kinds of defective behaviour.

5.11 Closing remarks

In practice, the restrictions placed on the consumption of environmental goods take two forms: regulation by the state and private appropriation. Hardin gives an example of the latter which is not historically correct: the enclosure of the commons by the landed gentry was designed to protect the land against overgrazing. Experimental research has led to the proposal of various solutions of this kind. According to Acheson (1975), dividing oyster banks into separate sections, each of which may be fished by only one fisherman, would greatly improve the management of the banks. Each man would concern himself only with his own section, and this would help to prevent the depletion of the banks. Without any doubt, game theory and the social dilemma approach is a very powerful tool in analysing environmental problems and formulating solutions. A strong warning however, should be put forward, as there is a strong tendency of misusing the approach. Two tendencies appear to be particularly dangerous, both connected with the obvious strong illustrative and educational character of game theory. The first is the tendency toward oversimplification leading the use of metaphors as a basis for policy prescriptions. Policies based on metaphors can be harmful, as they often lead to results substantially different from those presumed to be likely, as Ostrom warns rightly (1990, p.23). Part of this tendency toward simplification is the large emphasis in the environmental management literature on the significance of the prisoners' dilemma (section 9.8). The second tendency of oversimplification is the search for 'the only way' to solve a social dilemma, whether that way is found in the powerfull state or in strict privitization (Ostrom, 1990, p.14).

No matter whether the state or private individuals are responsible for preserving collective goods, the restriction of free consumption does not automatically result in better management. Aside from the fact that many environmental goods are not exclusive, and cannot be appropriated by anyone, it must be said that some goods entrusted to private individuals are destroyed at the same rate as they would have been by society. In any case, private management does not prevent the transfer of disadvantages to future generations. On the contrary: pollution of soil is often a result of many years of use by the owners, as in the case of factories. It is only after their departure that the problem comes to light, and by then it has already been transferred to the next users.

This transfer of the costs of environmental damage is one of the most difficult problems now. Many of the social dilemmas which now present themselves will ultimately affect people who have not yet been born and thus have no say in the decisions. Because they are not in a position to defend their rights, they must rely on today's 'caretakers'. Often the environment has no natural lobbies or pressure groups, which is seen as a fundamental problem by environmental organizations.

In the end, all these considerations point to the state as the guardian of collective interests, and the protector of the environment. Game theory is a powerful tool that can be used to enhance our understanding of the role of government and law in social life, but it should always be used along with other approaches (Baird *et al.*, 1995).

As there are many different kinds of dilemmas, an all-powerful, technocratic state does not always represent the best solution. In fact, the reverse is sometimes true. In the search for a solution to the many social dilemmas confronting us, the state has a role to play, an issue that will be discussed in chapter 9. There are no recipes that can simply be applied to all environmental issues. As the individual preferences often show a large variety and the types of dilemma's concerning environmental goods are so different, it certainly must be a role characterized by well-considered and balanced decision-making.

Chapter 6

Environmental attitudes and behaviour

6.1 Introduction

By definition (section 1.2) the causes of environmental degeneration are to be found within society, not in nature itself. Environmental issues are created by the human society. At the same time, an environmental problem only exists if society or some part of it has recognized it as a problem. This means that a natural phenomenon is only an environmental problem when:
a: it is caused by some kind of human activity;
b: it is recognized by human beings and judged to be a problematic phenomenon.
Both aspects, causes as well as responses to environmental problems, are the object of study within the social and behavioural sciences. In two chapters we will concentrate on activities and responses at the level of individuals. In chapter 7 we will discuss the reactive side: the significance of environmental changes for individuals, the way these changes are perceived and what kind of responses they may induce. In this sixth chapter we will concentrate on the active side: individual behaviour, motives, backgrounds and the relevance of it for environmental damage.
Three levels of personal tendencies are distinguished in this chapter. We are referring to *values*, when we are talking about guiding principles in the life of individuals or groups (Rokeach, 1973). Values have to be distinguished from attitudes, which are psychological tendencies expressed by evaluating a particular object with some degree of favour or disfavour. If the object is a wide category of phenomena, such as the environment, the term *general attitude* is used. An example of a general attitude discussed in this chapter is the attitude toward the environment ('environmental awareness' or 'environmental concern'). However, it will appear that in the literature this general attitude is sometimes called a value. We will use the term *attitude* when the object is one particular kind of behaviour, separating household waste for example. *Behaviour* we will define as all forms of doing, refraining or tolerating, that are linked to any subjective valuation by the individual.

6.2 Relevant behavioural patterns

We will start with a short description of long term trends that we regard as causes of growing environmental stress. The three structural trends described in section 6.2.1 are linked to economic growth and technological development. Both are macro trends, but they are a result of the individual behaviour of numerous persons. Economic behaviour, for example the way we have organized our production and the level of material consumption we aim for, has tremendous consequences for the environment. Hence behaviour is widely recognized as a fundamental cause of ecological degradation. Assuming that someone's behaviour is generally part of a particular behavioural pattern, based on values and attitudes, changes in values and attitudes are often recommended as a solution for the environmental crises. A 'real solution' could only be achieved by a radical change in mentality, followed by a changing of critical behaviour on a population-wide scale (Maloney and Ward, 1973). The expression of this view can also be found in many reports and documents issued by the environmental movement in the seventies. In this chapter we will examine behaviour as a cause of environmental stress, attitudes as cause of behaviour, and attitudes as a result of environmental values.

If environmental problems are basically caused by a succession of environmentally damaging acts by individuals, the solution has to be found in behavioural change. However, while this 'solution' has been widely recognised, the key question appears to be how to achieve such a radical change in behaviour. Which means are appropriate for stimulating a new kind of behaviour and who is supposed to use these means? The question could not be answered, but the notion of the behavioural background of environmental degradation stimulated research on environmentally relevant behaviour and attitudes.

6.2.1 Significant developments

An electricity company decides to use coal for the generation of electricity, instead of much cleaner natural gas. A ship dumps oil at sea. An agricultural company uses pesticides that in the end will make the groundwater unsuitable for drinking water. An army battalion destroys forest and heath. These are all examples of environmental damage caused by human behaviour. However, the damage is done within the framework of an organization in which these people are participating. Their behaviour is not directed at fulfilling their own needs, but it is a personal task aimed at achieving the goals of the organization. It is behaviour with environmental impact within a collective pattern, so called institutional behaviour.

There are many behavioural operations with environmental impact which people carry out on their own initiative, either individually or as a part of a family or a group. An individual will be influenced in his actions, although he is still controlling his behaviour himself. The informal groups may influence his behavioural intentions by means of social norms. Only in a part of all actions are individuals the only initiators of their behaviour and then the behaviour is determined by a person's own motives and considerations.

In spite of these three fundamental different backgrounds of behaviour, institutions, social norms and personal motives, we can recognize some general behavioural trends in society that have severe environmental consequences.

Over-population and *population growth* is one of the fundamental developments causing environmental problems (Ehrlich and Ehrlich, 1990; Shaw, 1992). The significance of this trend has been discussed in section 1.4. Population growth is determined by the number of births and the number of years people survive. A determined effort to limit the life span for the purpose of environmental preservation is not acceptable in today's society. On the other hand, everyone is more or less free to limit his or her reproduction voluntarily for environmental reasons.

Industrialized countries have moved from an equilibrium with high birth and mortality rates to an equilibrium of low birth and mortality rates, a development called 'demographic transition'. The emancipation of elderly people who were no longer dependant of their progeny for their living, and the emancipation of women fulfilling more tasks than giving birth to children and raising them, strongly contributed to a decrease in birth rates. Beside, the number of undesired pregnancies and births has decreased by means of improved contraceptives and education. The limitation of population growth is a healthy trend from an environmental viewpoint. It is based on a large number of behavioural operations that scarcely have a link to environmental considerations as such.

A second fundamental trend is the *quality and quantity of consumption* (sections 1.4.1-2). The amount of consumption is largely linked to the increase of welfare and therefore to economic growth. The strong relationship between economic growth and the growth in energy demand is shown in chapter 1 (fig. 1.1). The increase in energy consumption takes place directly in dwellings and in personal transport, and indirectly by means of consumption of products and services. Higher *per capita* energy consumption is the outcome of changes in consumption behaviour. We are choosing higher temperatures in our homes, we are heating more rooms, we have more energy consuming appliances, there is more motorized mobility. A large part of this consumption is individually motivated behaviour, but another part is institutional behaviour. For example, the installation of central heating systems in many countries in the last three decades, has been stimulated by building regulations. Because in houses with central heating a substantially higher amount of energy is used, the *per capita* energy consumption for heating purposes has grown. In general *per capita* consumption of products and services has grown strongly. In the production process large quantities of energy are used, so the products and services contain a certain 'energy volume'. Growing consumption is directly and indirectly causing higher energy use *per capita* (section 6.4).

The third significant trend is the geographical *spacing of human activities*. The trend has two components: the spacing of living and the growth of mobility.

Individuals only have limited control over their housing. There is a limited freedom of choice of where to live and in what type of accommodation, in particular for the lower income groups. Wealthy people generally have more options and are therefore using more space. The higher the income, the larger the possibilities for people to take their own decisions. There is a trend towards more space *per capita*, which might be called 'dilution of living'. The number of persons per household is decreasing in most countries, which also implies a larger number of households. The result is that on average all individuals are occupying more room and less space is available for functions other than dwellings. Firstly the tradition of different generations living together has decreased and secondly of couples living together has decreased. The number of one and two person-households has grown enormously in all western industrialized countries. The same emancipating processes of women and the elderly that create the environmentally healthy trend of a smaller population growth, reinforce the growth in space needed for housing. The increase in spacing is supporting growth in mobility as well. Motorized personal transport contributes to many environmental problems (OECD, 1995). In this respect the individual behaviour is largely determined by trends in society over which the individual does not have personal control. In particular distances between house and work are a result of collective decisions. Physical planning and urban development determine the distances and are also determining the possibility of choosing between modes of transport. In many cases urban planning is limiting the possibilities for environmentally sound ways of travelling, like walking or cycling. Nevertheless the distance between home and work is partly also a matter of personal choice.

The concept of attitudes is generally used for studying the motives of particular types of acts. The trends that have been discussed are all based on individual behaviour in which specific patterns can be distinguished. Every person, to a certain degree, has a stable behavioural pattern in accordance with personal values, social norms and his or her personal living conditions. General patterns of behaviour are usually called 'life-styles' (Schipper *et al.*, 1989). Life-styles may vary when individuals or households are compared, but these variations are only within a certain range that is culturally defined. At this aggregated level it is widely recognized that the changes in general behavioural patterns or life-styles are mainly responsible for growing environmental stress, in particular the growth in energy consumption (Schipper, 1996). An example of the significance of life-styles at the individual level will be discussed in section 6.8.3.

The conditions for individual behaviour are set by society and basic dominant values (as discussed in chapter 4) are guiding individual behaviour. According to the supposed impact of basic values on day to day behaviour (Rokeach, 1973), much of the research was directed at the development of new values linked to environmental issues (Dunlap and Van Liere, 1978). For this purpose the concepts of *environmental awareness* and *environmental concern* were

introduced, which are used for two purposes.
The first objective for the introduction of this concept is to establish the development of a value based on environmentally motivated concerns. These concerns would constitute a new developing world view, called environmentalism (Milbrath, 1984), a new environmental paradigm (Catton and Dunlap, 1978), an ecological paradigm (Catton and Dunlap, 1980), or an ecological world view (Blaikie, 1992). These values will be further discussed in section 6.5. The new values will become manifest in environmental attitudes that should assess the extent to which people who are concerned with the environment differ fundamentally from the less concerned. As a logical consequence the second objective for the introduction of the concept of environmental awareness is to establish causal relations between concrete behaviour and attitudes based on environmental concern. In both cases research is aimed at uni-dimensional concepts, such as environmentally sound activities versus activities that are causing environmental stress, or attitudes supporting environmental values against attitudes disregarding the environment. However, as empirical attitude-behaviour studies point out, in day-to-day life the division of attitudes as well as behaviour between the simple antitheses of environmentally sound and -unsound is precarious.

6.2.2 Consumption behaviour

The trends discussed in behavioural patterns are responsible for increased environmental problems, because consumption growth is causing an increase in the use of environmental functions. However, within these trends large differences in individual consumption behaviour exist.
In all purchases and uses of products or services a number of decisions are taken. Do we purchase a product or don't we? If we do, what do we purchase? And when we have bought something, how and how often do we use it? All choices are directly or indirectly related to environmental impact. Because consumption is motivated by the fulfilment of personal needs the motives for our consumptive decisions are usually not based primarily on environmental values. However, if they are, and we renounce the purchase or use of a product for environmental reasons, how do we spend the money that has been saved? That is the next consumer decision with environmental consequences, because even if we decide not to spend our budget and leave it on our account, the bank will use it for investments and consumer credits.
The choices made when purchasing and using products or services are shown in figure 6.1. On all paths in the diagram a decision is made that is partly determining the final environmental impact. The relative environmental impact is shown by positive or negative signs. The impact is relative, because it must be compared with alternative choices in all cases. The decision for public transport is a positive one compared with driving a car, but it is negative

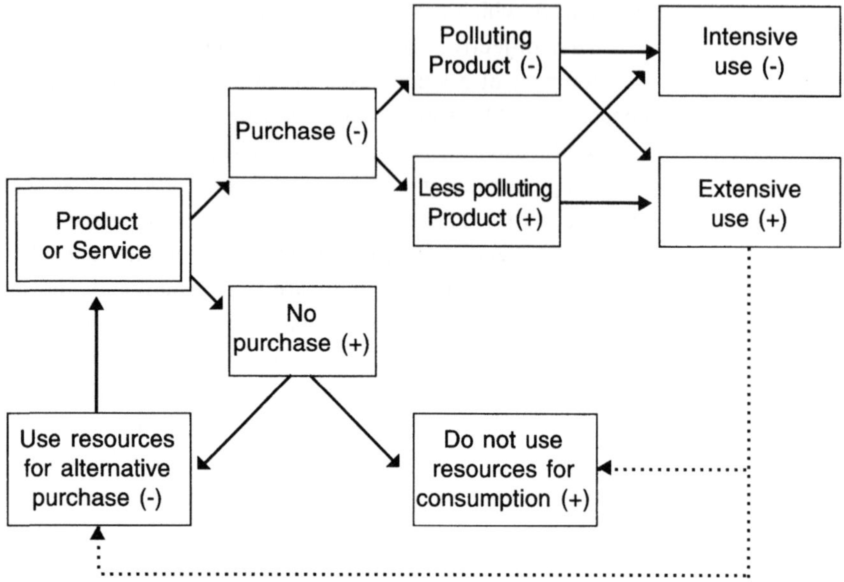

Figure 6.1 Scheme of choices in consumer spending; on each decision relative environmental impact is indicated

compared with cycling. Using polluting products sparingly, for example batteries, is only 'environmentally sound' compared with more extensive use of the same product. When looking at the impact of the product its whole life cycle has to be taken into account, not only the impact of the production, distribution and use, but the effects of the product in the phase of waste as well.

Someone is buying a tumbler for drying the laundry for example. This equipment is causing environmental stress because of the materials and energy used for production, the energy used for transportation and in particular because of the energy it needs for domestic use. Nevertheless, there is a variety of models from highly energy efficient to machines that waste energy. When the tumbler is purchased it may be used intensively, but it may also be used selectively, for example only when the weather does not allow washing to be dried out of doors. Anyone who does not buy a dryer is saving money: the price of the purchase and the costs saved on the energy bill. This leaves the question: for what purpose this money will be used. When it is used for an air trip the result, looking at environmental consequences, is most likely negative.

6.3 What is environmentally sound behaviour?

One way or another within all our daily behaviour there are hardly any actions that do not result in some environmental stress. Behaviour that is partially motivated by the consequences of the acts for the environment, is mostly categorized as 'ecologically sound' or 'environmentally sound'. Looking at figure 6.1 it is obvious that these terms are not as unambiguous as they are often presented. In this section we will attempt to type behaviour according to its environmental relevance. In the literature this kind of behaviour is often called 'environmental behaviour' (Ester and Mandemaker, 1992). For the time being we will use this term, but will be more specific about this subject later because of fundamental problems linked to the term 'environmental behaviour'.

6.3.1 A definition of behaviour

First we have to consider what exactly should be called behaviour. Here we will use a sociological definition: behaviour includes all forms of doing, refraining or tolerating, that are linked to any subjective valuation by the individual (Weber, 1968, p.1). In this sense all acts with consequences for the physical environment may be called 'environmental behaviour', acts with positive consequences as well as the ones causing negative impact. However, it appears to be difficult to describe positive acts as 'environmentally sound' behaviour, or 'ecological behaviour' (Nelissen and Scheepers, 1993 p.76). 'Environmentally sound' is always a relative concept as pointed out for consumper decisions in figure 6.1. What is called positive environmental behaviour is dependant on what may be considered as usual or 'ordinary behaviour'. Whether or not an act may be considered positive is relative to the environmental consequences of the 'ordinary' alternative.

What should be considered 'environmentally sound' behaviour?

An Englishman wants to spend a weekend in Paris. He is considering three options for travelling. Going by car and ferry is negative environmental behaviour. A positive alternative is using public transport and a train through the 'chunnel' is now available.
Taking the train to Paris is usually considered positive environmental behaviour. However, a trip to Paris as such is creating environmental impact anyhow, even when the train is used. Compared with staying at home or a weekend in England it may be called negative environmental behaviour. Anyone who is travelling a lot and makes long motorized trips, even when using public transport, has a partially environment burdening life-style. The positive option is staying at home. The label 'positive' linked to environmental behaviour is in many ways dependant on the alternative way to act. Even the trip to Paris by car may be called positive, as the alternative for many people would be an air trip, or some would call it positive as the trip is made in a car 'pooled' with others.

In common language many types of behaviour that are actually a burden to the environment, are called 'environmentally sound'. Using public transport is one of them, as well as driving an energy efficient car that is using unleaded fuel, or heating the home with a highly efficient central heating boiler, building a wind turbine, using a gas heated boiler instead of an electric boiler. Al these acts are called 'environmentally sound' because there is an ordinary alternative that is causing more environmental impact. However, these types of behaviour are still a burden to the environment and for many of them there are even better simple alternatives.

6.3.2 Typology of 'environmental behaviour'

In the whole range of behaviour with a certain relevance to the environment five types may be distinguished. For this typology two criteria are used: the physical effect of the behaviour (whether or not it affects the environment) and the social meaning of the behaviour (negative versus positive).

Type I: behaviour burdening the environment.

This is a category of behaviour with environmental consequences that are easy to avoid. The impact may easily be reduced by choosing alternatives that are more environmentally sound (types II and III).

Type II: behaviour sparing the environment.

A category of behaviour consisting of alternatives for type I behaviour. There are two possibilities. First: behaviour that is causing less environmental trouble that replaces behaviour with severe environmental consequences. For example: going to work by public transport instead of a car. Second: operations that are limiting environmental impact that are an extension of existing behaviour. If a clear choice is made for these alternatives for reasons linked to the avoidance of environmental impact, behaviour of this type may be called 'environmentally sound'. An example is separating household waste, which is not limiting the amount of waste but creating an opportunity to reduce landfill and incineration.

Type III: refraining behaviour that is a burden to the environment.

This is the largest category, but also invisible in most cases because it includes acts with environmental impact that are *not conducted.* This is the most important environmentally sound category of behaviour. Nevertheless it is hardly recognized because there is a whole range of motives for it and many of these motives are not linked to any kind of ecological consideration. Within this category a distinction should be made between *type IIIa*, environmentally motivated behaviour, and *type IIIb*, without environmental motives.

An example is a wintersports vacation. Anyone who is not skiing is conducting environmentally sound behaviour, at least when the money is not spent on another trip with an even greater environmental impact. The motive for staying home or planning another vacation trip may be ecologically based (type IIIa), but another obvious motive is lack of money needed for an expensive wintersports vacation. Yet other reasons for staying home are that not everybody likes skiing, nor is physically able to do so (type IIIb).

Type IV: behaviour directed at environmental improvements.
Any kind of behaviour conducted for recovery of environmental damage or the prevention of it belongs to this category. Examples are participation in environmental actions or the support of environmental organizations (see section 8.2), or participation in pollution prevention activities such as re-use of products or the use of recycled products. A more direct type which is in most cases institutional behaviour, is cleaning polluted soil, planting trees, recovering damaged ecosystems by re-introducing species. This type IV behaviour is easily recognized as environmentally sound as the motives are predominantly ecological.
Type V: frustrating environmentally sound behaviour.
This is a category of mostly institutional behaviour without direct environmental impact. It is directed at maintaining or extending the possibilities for environmentally unsound behavioural alternatives. The way many products are offered, in package which consumers often consider oversized and unnecessary, is an example. Another striking example is the installation in buildings of appliances which cannot be switched off, for instance lighting, ventilation, air-conditioners or automatic awnings.
The categorization of behaviour as negative against positive, as environmentally sound against environmentally unsound, is far too simple. In particular 'positive environmental behaviour' is divisible into several types of behaviour of a very diverse nature. The division of behaviour to environmentally sound- or unsound does not take the subjective meaning of an act into account, which is a significant part of the definition of behaviour. Most acts which have to be considered 'environmental behaviour' because of relevant environmental consequences, are not performed taking these consequences into account. Even if the environmental consequences of behaviour are actually taken into account, the qualification of behaviour as environmentally sound- or unsound cannot be uni-dimensional either, because environmental quality is not uni-dimensional. For example fox hunting will be qualified as environmentally unsound by many people, but some environmental organizations accept it as a measure necessary for the maintenance of other species in natural reserves. However, in most categories of behaviour there is simply no awareness of the damage done to the environment. Hence 'environmental behaviour' does not make sense as a scientific concept.

6.4 Environmental damage as a side-effect

Empirical research on behaviour which is relevant for the environment mostly shows hardly any correlation between concrete types of behaviour. The reason is that there are no social or psychological criteria primarily determining whether or not behaviour is relevant to the environment. The criteria for characterization of behaviour are of a physical nature, lying beyond social or psychological categories.

Because there is no common human origin, behaviour relevant for the environment cannot be described within one dimension. There are no common origins, only common results (environmental damage) and in general these results are only side effects of human behaviour. In most cases these affect crucial functions within society, but this social impact is not a motive for the behaviour. Environmentally damaging behaviour is not aimed at environmental damage, nor at the social impact of it. The purpose of a car driver is to reach his or her destination, not to pollute the air. The number of acts aimed at environmental damage is very limited. Only in some cases of malicious destructive acts of warfare, such as the use of the defoliant 'agent orange' in Vietnam, or the firing of oil wells in Kuwait by Iraq, is environmental damage not a side effect.

Behaviour directed at environmental improvements (type IV) and ecologically motivated avoidence of behaviour with environmental impact (type IIIa) are the only types which can be classified in a socially defined category. Participation in environmental actions and separation of household waste are both motivated by ecological considerations. Most behaviour that is usually considered as environmentally sound, is of the not ecologically motivated type III. In many cases people are hardly aware of the fact that they are conducting environmentally sound behaviour by refraining from behaviour with environmental impact, simply because they have quite different motives. Limited financial resources are the main reason for ecologically sound decisions in the case of consumer behaviour. As a consequence people are embarking on new environmentally damaging activities as soon as their financial budget increases.

An illustration is shown in figure 6.2. It shows an estimation of total energy consumption of households. Total consumption is not only direct consumption for heating, power and lighting in the house. For consumper goods like refrigerators the phase of household use determines the largest part of the total energy consumption, but for many products the production phase is more important. All goods from food to housing, from clothes to toys, and from sending a letter to taking a vacation, need energy for production, transportation, distribution and waste handling. The energy volume of these products is the source of indirect energy consumption which is incorporated in the estimations of figure 6.2a. In figure 6.2b the relation between income and energy consumption is shown in a different way. The indirect energy intensity of consumption expenditure is practically constant over all categories, as the direct intensity is slightly decreasing.

On average the total use of energy is directly linked to household income. A higher income is followed by an increased level of direct and indirect energy consumption. The reason for it can be found in figure 6.1. Decisions about consumption can be made with or without environmental considerations in mind, but the financial budget of the household will be used for consumption in one way or another. As all goods are directly or indirectly linked to energy use, for all households a large financial budget will be linked to a high level of energy

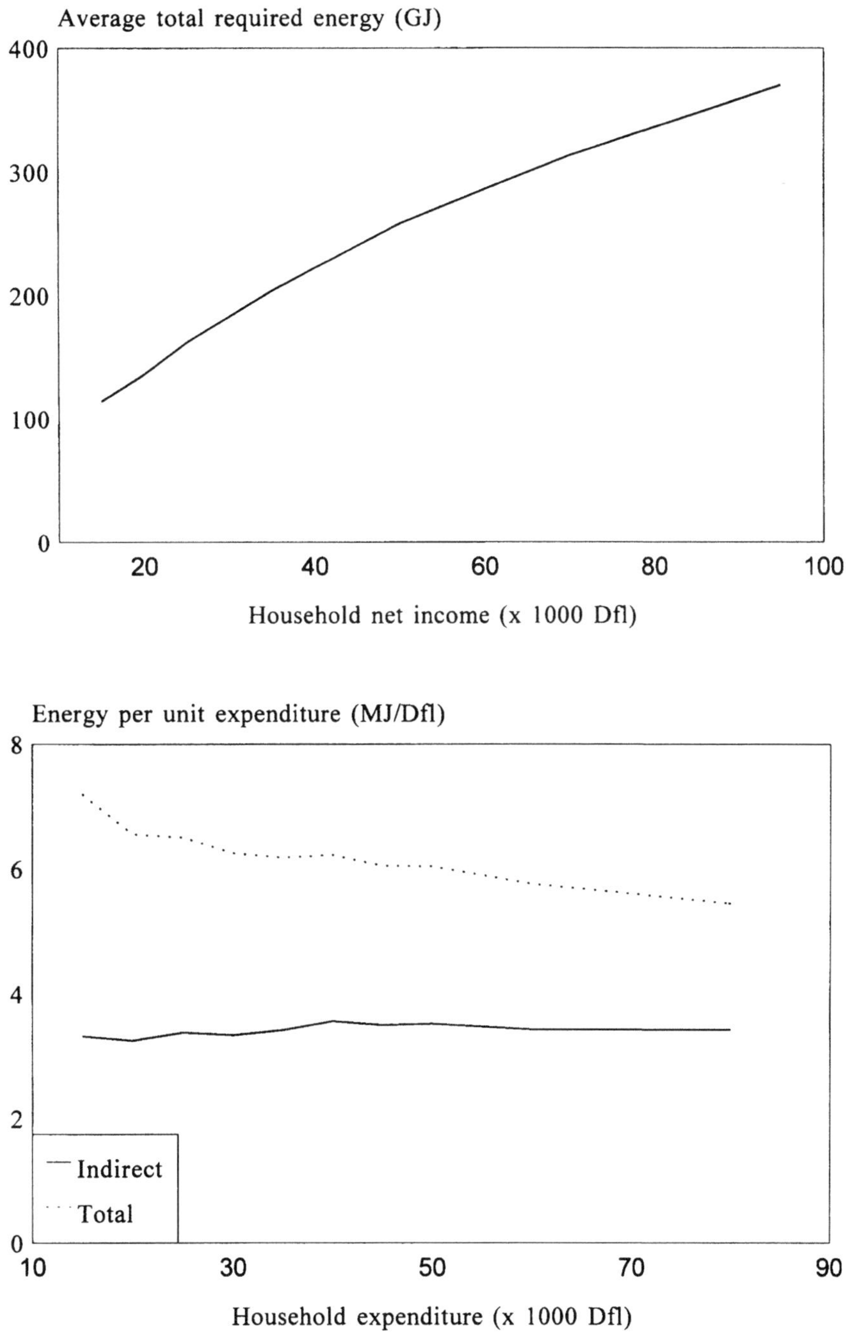

Figure 6.2 (a) Total energy consumption related to net household income. (b) Direct and indirect energy consumption per unit of expenditure (Source: Biesiot et al. 1995; also reported in Vringer and Blok, 1996).

use. Figure 6.2 is based on a comparison of income groups. However growing average welfare could be compensated by trends of increasing efficiency in the use of energy and materials, so called *dematerialization* (Bernardi and Galli, 1993; Von Weizsäcker *et al.*, 1995). At present these trends are not strong enough, so income growth actually implicates more direct and indirect energy consumption. After describing the trends in waste production and in energy and material use, Warnick *et al.* (1996, p.194) state: 'we see no signs of net dematerialization at the level of the consumer or saturation of individual material wants.'

Of course the figures shown in 6.2 are averages. At all levels of income there is variance in which differences between households become apparent. The differences are determined by other factors of life-style, such as the use of the time-budget of households and the phase in life-cycle of households (section 6.8). Within this part of variance a more or less sustainable life-style may also cause some differences. High energy life-styles with preference for consumption of many high energy volume goods and services like a jacuzzi, an air conditioner, cars, air trips and so on, may be contrasted with relatively low energy life-styles, including goods and services like books, most forms of education, gardening. Even in the latter case it may be hard to recognise the energy volume of goods for people that actually try to live in a sustainable manner. Like all other types of environmental burden the energy used for production, transportation, use and discarding is a side effect that in many cases is hardly visible for the consumer.

6.5 Environmental awareness

Data on public perceptions of the environmental issues are interesting as an indicator of the attention paid to the environment within society. Public opinion time series-data show changing attention and concern, cross-sectional research may indicate significant variations between countries or population groups. How can these variations be explained and can they be interpreted in terms of a growing significance of new environmental values? What is environmental awareness? Is there actually an increasing significance of it? And is this awareness founded in changing values?

6.5.1 Variations over time: the attention cycle

In many countries a decrease in public attention for environmental issues in the eighties compared to the seventies has been recognised. Such a decrease was forecasted by Downs in his description of the 'issue-attention cycle', a concept he developed to describe the fluctuating attention for issues of crucial importance and which he primarily used to explain the rise of public attention for ecological questions. In the issue-attention cycle, rooted in the nature of the problem as well as the way mass media operate, Downs distinguished five stages, almost always occurring in the same sequence (Downs, 1972, p.39).

I. The pre-problem stage:

A highly undesirable social problem exists but has not yet captured public attention. Only experts or interest groups have noticed the issue, and may be alarmed.

II. Alarm stage: discovery and euphoric enthusiasm.

For whatever reason, often as a result of a disaster or a series of dramatic events, the awareness of the public suddenly arises. It is discovery that causes alarm that will be accompanied by euphoric enthusiasm about the possibilities of solving the problem. Downs partially explained the linkage of alarm and enthusiasm with a typical American tradition of optimistically viewing most obstacles to social progress as external to the structure of society itself. Why the environmental issue came up in other countries cannot easily be explained.

III. Realization of the cost of significant progress.

The fact that the costs of solving the problem, not only in terms of money, will be much higher than expected damps the enthusiasm. There is an increasing recognition that the causes of the problem are also to be found in arrangements within society that are providing benefits to certain groups. In terms of social dilemmas (chapter 5), it is the understanding that distributional conflicts emerge while solving the problem because it has the characteristics of a zero-sum game (Thurow, 1981, p.120). For an issue like the environment it means that technical solutions to which all may contribute proportionally are far from sufficient and some parts of society will have to pay more than others.

IV. Gradual decline of public interest.

More and more people are starting to realize how difficult it is to solve the problem and how costly it will be to themselves. Some will feel threatened by it and start trivializing the problem (dissonance reduction, see also section 7.5.3), others just become discouraged or bored by the attention given to the problem, for example in the mass media. An important factor also is that total public attention can only be spread over a limited number of issues. Downs states that in this stage some other issue is usually entering the alarm stage.

V. Post problem stage.

The issue is out of focus, probably replaced by others. It 'moves into a prolonged limbo, a twilight realm of lesser attention or spasmodic recurrences of interest' (Downs, 1991, p.40). Note that the problems are not yet solved. They simply do not attract much attention any longer and the priorities lie elsewhere, which does not mean that absolutely nothing is done, as earlier stages might have resulted in surviving institutional arrangements that are still dealing with the issue.

6.5.2 Environment in the attention cycle

By definition the problem only emerges in the first stage of the attention cycle. The mere fact of any physical change does not constitute an environmental problem. It should be recognized within society and valued as an undesirable phenomenon. After its recognition the 'environment' as a general issue certainly reached the second stage of alarm, but for specific environmental problems the second stage may sometimes not be reached, simply because some barriers may prevent the issue either being noticed or being valued negatively. We will discuss this later in chapter 9, where the agenda building theory of the political process is presented.

In her book *Silent Spring* Rachel Carson described in 1962 the common denominator of environmental destruction, in particular the consequences of the use of DDT. 'We have subjected enormous numbers of people with these poisons, without their consent and often without their knowledge' (Carson, 1992, p.29). Her message caused a shock. Generally speaking the alarm stage in the US and most other countries in the western world occurred at the time Silent Spring was published.

Downs, while describing his attention cycle, stated that in 1972 the third stage was reached. At that moment publications like *Limits to Growth* (Meadows, 1972) marked a climax in the attention for the environmental issue. As a result of these reports, however, an awareness also arose of the huge efforts needed to prevent the problems and disasters forecasted in the future, which started a decline of public interest.

In an empirical attempt to test the hypothetical attention cycle, it was established that for all issue-areas included in the research, one specific decade could be assigned in which the attention peaked. In this analysis energy and natural resources were also included, and the attention paid for these issues even preceded environmental pollution as a hot issue (Peters and Hogwood, 1985, p.247). However, the level of public attention wanes in all cases, including all types of environmental issues, which supported Downs description of the cycle. Although he forecasted the decline of public interest, he also noted that the decline in the cases of the environment would be less and slower than for most other issues, because new environmental problems would frequently announce themselves and most should be considered as a threat to society as a whole, and not only as a minority problem (Downs, 1972, p.50).

Downs considered public attention as a favourable condition for significant measures, not as a phenomenon that would cause significant changes in patterns of behaviour. In his cycle we can hardly recognize a shift in fundamental values, because such a shift should last longer than a decade. So the question arises whether or not public attention may be called a good indicator of environmental awareness in the sense that it causes changes in behaviour.

6.5.3 Social-cultural factors and value orientations

In an analysis of relevance of several background variables Preisendörfer found that the difference between East- and West-Germany appeared to be the only factor with a significant and consistent impact on all types of environmental awareness (Preisendörfer, 1996). The sensitivity to environmental issues in the former German Democratic Republic lagged behind, compared to West-Germany. Obviously fifty years of separation as well as the persisting different social and economic situation in the two parts of the country, is causing a difference in attitudes. Such differences, although less striking, are also consistently found in research concentrating on factors like education, income, sex, age, and even ethnicity (Taylor, 1989). These established differences show a great consistency, which suggests that there is most likely a certain foundation for them. The consistency in the differences in environmental attitudes is by researchers usually linked to human values.

Values are guiding principles in the life of individuals or groups (Rokeach, 1973). Several attempts have been made to develop indicators of general environmental concern, such as a scale for ecological attitudes (Maloney and Ward, 1993), or the NEP-scales (New Environmental Paradigm; Dunlap and Van Liere, 1978). These indicators provided a rationale for linking environmental awareness with environmentally responsible behaviour. However, there is still an on-going discussion about the values and the changes in values that are responsible for environmental awareness (Kempton *et al.*, 1992). The environmentalist value, as reflected in the NEP-scales, should be distinguished from the dominant social paradigm, which emphasises economic growth, material welfare and progress (Dunlap and Van Liere, 1984; Milbrath, 1984).

Originally the existence of environmental values contrasting the dominant paradigm (human excemptionalism paradigm; Catton and Dunlap, 1978), was supposed to be a one-dimensional phenomenon. However, research with the NEP-scales soon tended to show a multidimensional structure of environmental values. Albrecht *et al.* (1982) empirically found a reflection of different values within a set of indicators for environmental values. They established empirically the existence of three dimensions: concerns about the 'balance of nature', concerns about the 'limits to growth', and a value reflecting the anthropocentric view of 'humanity over nature'. These findings were merely empirical and descriptive, without offering a clear understanding of the reasons why these dimensions should be distinguished. These findings are raising the question whether or not scales like the New Environmental Paradigm really reflect basic values, or are merely indicators of general environmental attitudes. The NEP-scales as indicator of a new paradigm, to be distinguished from a 'human exceptionalism paradigm' are often severely critisized, in particular by Buttel (1978; 1986). Although this was a discussion on the validity of NEP as a paradigm within sociological theory, the fundamental question raised by Buttel is also whether or not NEP is actually reflecting one of the basic values that are shaping the social force behind future social change (Humphrey an Buttel, 1982, p.227). Much of the research on environmental attitudes and behaviour is based on assumed rational behaviour. Researchers are using subjective expected utility models, or more specifically the expectancy-value model of the theory of reasoned action of Fishbein and Ajzen (section 6.6.1). Actually the assumed inherent motivation of people in these social-psychological models is an egotistical value orientation, such as Hardin emphasized as the only predominant motivation (see section 5.4.1). However, from our description of the impact of values on the behaviour in social dilemma situations, we know that different value orientations may be dominant for different individuals, and for different categories of behaviour as well.

When environmental values are based entirely on rational maximization of utility, they are finally founded on self-interest. Beside this utility-oriented rationality, often even narrowed to economic rationality, moral values can be significant in shaping environmental awareness (Stern *et al.*, 1986). Schwartz (1977) developed a theory for norm-activation of altruism, that was fruitful applied in research on environmentally relevant behaviour (Vining an Ebreo, 1992; Hopper and Nielsen, 1991). The findings of these investigations suggest that, at least for specific types of environmentally relevant behaviour, environmentalism may be linked to concerns for the welfare of other human beings. Ego-centric and altruistic value orientations will be present in all individuals, but in varying proportions. Environmental awareness is therefore a multi-dimensional phenomenon, flowing from either egoism or altruism, depending on the personality of the individual, and the type of environmentally relevant behaviour we are looking at. Examples of two particular actions which are dominated by very different value orientations are recycling behaviour (altruistic, type II) and using alternatives for car driving (egoistic, type IIIb). We will refer to these examples more often in the rest of this chapter. Stern *et al.* (1992) are using this differentiation in value orientations for explaining the consistently found significant differences in environmental awareness between men and women, which would be caused by gender-based socialization processes. For the purpose of explaining environmentalist values, they even distinguish between human based altruism and a type of non-egoistic behaviour 'motivated by biospheric values that extend beyond the human species' (Stern et. al 1992, p.326). This agrees with the findings of Albrecht *et al.* (1984), although the explanatory power of this biospheric morality for behavioural patterns is rather limited. This is considered to be a result of socialization as well: when children are exposed to the dominant social paradigm, there would be a drawback in this nature-oriented environmental values. Nature is apparently much stronger in the value structures of children. Szagun and Mesenholl (1993) for example, found higher levels of emotionally driven concern about nature in younger children.

6.5.4 The significance of general environmental attitudes

Although there is a wide recognition of the long term importance of environmental values, the impact on behaviour of these values or even of general

environmental attitudes is rather limited. In empirical research the indicators of general environmental awareness are only slightly correlated with actual behaviour and they have little predictive power. When we take notice of the fact that the environmental relevance of almost all types of behaviour is only an indirect and secondary motive, these results should not be surprising. Consumption behaviour is directed at fulfilling needs, whatever the origin of these needs may be. The impact on the environment is nothing more than a side effect in all cases, so it is only a secondary consideration as well.
Stern and Oskamp (1987) argued that general attitudes such as environmental concern deserved less attention in research directed at explaining environmentally relevant behaviour. For an improvement in the prediction of behaviour, more specific attitudes should be investigated (Vining and Ebreo, 1992). This will be discussed in the next section. However, because of socio-cultural differences in value orientations which are consistently found, general environmental concern still seems a valuable concept. Basic values as reflected in general environmental concern may not be good predictors of actual behaviour, but they can help to understand the backgrounds of more specific attitudes.

6.6 Behaviour specific attitudes

A frequently found result in attitude-behaviour studies is that the relationship between general environmental awareness and behaviour is weak, in particular because attitudes reflecting environmental concern are often not followed by decisions in favour of environmentally sound alternatives (Nelissen and Scheepers, 1993). A better attitude-behaviour consistency is found in studies concentrating on attitudes measured at the same level as the type of behaviour that is investigated. While studying whether or not consumers have the intention of buying an energy-efficient refrigerator, the environmental attitude will hardly predict behaviour, and neither will the attitude toward energy conservation. A consistent attitude measurement will take the refrigerator and its use as an attitude object.

6.6.1 The theory of reasoned action

Attitudes are psychological tendencies expressed by evaluating a particular object with some degree of favour or disfavour. Three components are part of attitudes (McGuire, 1985):

- ★ cognitions: characteristics *('attributes')* that may be attributed to an attitude object;
- ★ affects: the degree to which an individual appreciates an object;
- ★ the conative component: the behavioural aspect as a tendency or an intention to behave in a certain way or not.

In many studies on environmentally relevant behaviour, such as litter, waste recycling, energy conservation, mobility, participation in environmentally relevant decision-making etc., the *theory of reasoned action* (TRA) is used (Fishbein and Ajzen, 1975). In the attitude-behaviour part of this theory the attitudes are based on a combination of beliefs about and evaluations of attributes of the object. The object is the behaviour that should be predicted. However, the attitude is not influencing behaviour directly, it is only influencing a behavioural intention. In the other part of the model social norms, constructed with the beliefs and valuations of relevant others (as perceived by the individual itself), are influencing intentions as well. Behavioural intentions are causing behaviour, although several barriers may prevent intentions becoming actual behaviour.

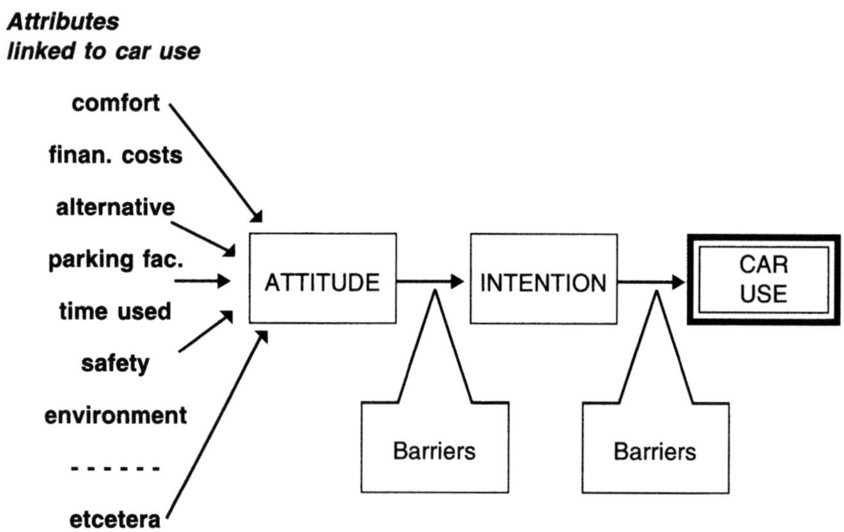

Figure 6.3 The relation between attitudes and car use according to the theory of reasoned action.

In figure 6.3 the attitudinal part of the theoretical model is presented in terms of car use, a type of behaviour with great relevance for the environment. The objects are presented in general categories, but they can be made more specific. For example financial costs can be divided into regular taxes, fuel prices, costs of maintenance, the price of buying a car etc, whereas environmental burden might be divided into air pollution, noise, using space, using fuel resources etc. Let us take an example on an attribute that is contributing to the car driving attitude. A potential car user attributes some time needed to find a parking place to his trip: a 'belief'. The contribution to the attitude of the estimated time needed is also dependant on an 'affect': the evaluation of this attribute, as time is not equally important for all. In the end the combination of beliefs and affects of all attributes that are perceived by the individual as significant regarding the decision to use a car, will contribute to a behavioural intention, while other attributes are not perceived as significant and will therefore have no impact on the attitude. In general, studies using the TRA are much more effective in predicting and explaining behaviour. The conclusions of this type of attitude-behaviour studies can be summarized:

★ There is only a direct causal relation between attitude and behavioural intention, not between attitude and actual behaviour.
★ There may be a relation between intentions and actual behaviour, but in between several barriers may exist.
★ Attitudes are based on all aspects of the behaviour that are considered important in the person's own perception, which social psychologists call *salience* (Van der Pligt *et al.*, 1988).
★ Attitudes only explain a substantial part of intentions if we look at attitudes linked to the specific acts we want to explain; it means that there is hardly a connection between the general 'environmental attitude' and a concrete decision, in particular not when the environmental side effects of a decision are not salient beliefs (see section 6.4).

The fact that attitudes as defined and measured in the TRA studies are predicting behaviour better than general attitudes, of course is one of the reasons of the popularity of the TRA model. However, this is most likely not the only reason. All governmental policies have certain presuppositions. Measures and instruments for policy implementation used by governments and public bodies are based on conceptions about society. In political science these conceptions are called *policy belief systems*. In most policy fields the basic assumption about people is that they act in accordance with the theory of reasoned action (Sabatier, 1987 p.665; also see section 9.4.4). A large amount of, mostly contract, research commissioned by governmental bodies, is carried out while using the theory of reasoned action with its rational choice background. For an illustration we will look again at car use, a type of behaviour that fits this assumption rather well.

6.6.2 An example of TRA research

For most kinds of behaviour with serious environmental consequences the current policy beliefs are that the fundamental origin of environmental problems has to be found within the mentality of the people and not within failing government policy itself. Therefore authorities show a tendency to start public opinion campaigns, sometimes using mass media, sometimes using more direct means of communication. The premiss of this kind of governmental actions is that influencing environmental attitudes is an instrument for behavioural change, in particular behaviour with environmental impact. So when a campaign is to be effective, we should know what the motives for car use are. Therefore we look at the salience of attributes in the case of car use. In a Dutch study by Tertoolen (1994) we find an answer to the question why there is no positive relationship between environmental awareness and abandoning car use. The attributes linked to the environment are hardly significant for anyone and the same holds true for safety aspects. In decreasing significance the order of attributes for choosing means of transportation is: perceived use of time, then, very much less significant already, perceived independance; comfort; health; financial costs; environment; safety. What may be expected from campaigns that try to change peoples attitudes using environmental or safety arguments? There is hardly anything to be expected from it looking at attitude change. There is even less reason for expectations as far as intentions and actual car use are concerned. Trying to change peoples beliefs about the environmental consequences is rather ineffective, as almost nobody considers it as a significant aspect when deciding how to travel.

Another result is that changing the financial costs will have only a slight effect. The order of significance of attributes shows that the perceived price of mobility is not the most salient attribute either. As a matter of fact the same study shows that people are not making a trade-off between facts about the salient attributes, they decide with subjective perceptions which are sometimes far away from the facts. They rely upon 'bounded rationality', which generally results in imperfect results according to the rationality assumptions of classical economists (Simon, 1987). Consumers generally lack the expertise to judge what consequences are linked to their behaviour, which is obviously the case for car driving, because the financial costs of car driving are greatly underestimated.

Financial costs appear to be less important than aspects of time, like speed and delays, and also less important than perceived independence. The most effective way to influence the choice for car driving is changing the perceptions about the time needed for a trip. However, this is hardly possible with any kind of persuasion, so publicity campaigns will not be effective. The real time needed for a car trip and the feelings of independence should be changed, for car use and also for the alternative means of travelling. This would mean that creating physical barriers and regulations limiting the range of behaviour are the way to change behaviour. To change time-perceptions measures such as limiting parking places, closing down streets for cars, speed limitations etc. would help, as on the other hand measures such as better public transport, establishing cycle routes, separate cucle lanes and walking routes could improve the time-consuming image of these alternatives.

6.7 Limited value of the TRA

6.7.1 Different types of behaviour, different values

The example of car driving is probably a good one to show the validity of the

TRA. Car driving is a type of behaviour that is highly determined by the more ego-centrist utility concept. However, in recent literature the validity of the TRA, at least for some types of behaviour is questioned, for example by Thogersen (1996, p.549) who does not believe that expected utility models 'can be fruitfully applied to all the diverse kinds of decisions and activities with which people cope.'

Why are Britons so bad at recycling?

Be honest. What will you do with this newspaper when you have finished reading it? Shove it in the dustbin or put it to one side in a special newspaper box? And what will happen to that empty bottle of cabernet after your Sunday lunch? If you're planning to visit a bottle bank, you are in a minority. Only about one in three people regularly recycles rubbish and overall the recycling of domestic waste in Britain hovers miserably around the 5 per cent mark, against a government target of 25 per cent (and a figure of 30 per cent in Germany).
It's not as if it's difficult: many supermarkets provide facilities. So why are we doing so bad at recycling? Perhaps it is the suspicion that we will be wasting our time. Even those who sort paper from plastic, Coke cans from champagne corks (yes, even these can be recycled) must sometimes wonder whether it is worthwhile. Stories of paper mountains dumped into landfills, even during a crisis in the world paper market, only feed suspicions.

Recycling makes us feel good, but does it any good? What happens to our rubbish after we have lugged it to the bank?...

The Times, March 9, 1996.

In section 6.5.3 we discussed the multi-dimensional character of environmental values. There the egotistic value orientation was distinguished from the value orientation toward general human welfare, a more altruistic position. Thogersen, among others, argues that for certain kinds of behaviour the altruistic dimension is more important than the dimension which is reflecting subjective utility. In an overview of the literature on separating waste and cooperation in recycling programmes, he concludes that participation is better explained by the altruistic dimension of environmental values than by TRA type attitudes. This is an important conclusion, because it has significant policy consequences. Research based on the TRA will suggest that for an improvement in participation in recycling the private side-benefits should be increased. This can be done by tariffs of any kind or by financial rewards (Allen *et al.*, 1993). Hopper and Nielsen (1991) found that attitudes based on altruistic environmental values were predicting recycling behaviour much better. Adding incentives directed at personal utility may undermine the tendency for participation in waste separation and recycling. 'Incentive schemes that reward proper and punish improper behaviour may reduce the feeling of obligation' (Thogersen, 1996, p.551). This may be a statement that is hard to understand by an economist, however it is not unprecedented. For example in the 'market' of blood donors: commercial blood markets appear to be less efficient economically and administratively, more expesive and less safe (blood quality) than voluntary donor systems (Titmuss, 1970, p.231).

6.7.2 Behavioural control

Recently the picture of concrete behaviour only explained by act-specific attitudes and social norms, has been refined as well (Manstead and Parker, 1995). Ajzen's (1991) modified model for attitude-behaviour relations, the theory of planned behaviour, takes perceived behavioural control into consideration. This is an important modification, because in the field of environmentally relevant behaviour this behavioural control is actually limited in many cases, because there are no behavioural alternatives available.

In the TRA limiting factors are considered as barriers between intentions and behaviour. Often these barriers are physical, in many cases they are determined by institutional behaviour and collective decisions in the past. Looking at the same examples we used so far: obviously separating waste and participation in recycling is only possible when there is a separate waste collection system. In some cases people started separating specific waste components on a voluntary basis, for example for the purpose of paper recycling. However, for components like green waste (kitchen, vegetable and garden waste), glass and aluminium cans, separate collection was started with a political decision to experiment with collection systems. For the example of car driving, the availability of alternative ways of travelling is a factor that is determining behaviour for a large part.

Physical barriers or institutional barriers may limit the control of the individual over his own behaviour. The point that is added in the theory of planned behaviour, is the psychological factor *perceived behaviour control*, which is related to actual barriers, but is determined by subjective perception as well. Before an individual develops an intention for a specific action, a certain feeling that the purpose of the action lies within the personal control (internal *locus of control;* Lefcourt, 1982) should exist. Based on the concept of *self efficacy* (Bandura, 1982), Ajzen's theory incorporates perceived behaviour control in the model as a factor shaping behavioural intentions. In the example of car use, the perceived behavioural control is for a large part dependant on the perceived availability of alternatives, such as public transport. Trains and buses, even when actually available, are often not considered to be satisfactory alternatives. The problem is that most behaviour is linked to other activities that have to be changed as well when an alternative that is environmentally more sound is chosen. Most environmentally relevant behaviour is part of daily life that is determined by habits and life-style patterns (section 6.8).

6.7.3 *Habits*

A third aspect of limited validity of attitude-behaviour relations is the impact of habits. Macey and Brown (1983) investigated repetitive behaviour in the case of household energy conservation and they showed that past experience with specific behaviour is often a better predictor of behaviour than attitudes. Habits can prevent the establishment of environmentally more sound behaviour.
These results suggest that the attitude formation might as well be influenced by (experience with previous) behaviour, as by an inversion in the causality of attitudes and behaviour: attitudes that are determined by behaviour ('construction by aspects', McGuire, 1985 p.243). Beliefs about several aspects may follow behaviour, instead of causing it. This is likely to be rather common, in particular for repetitive behaviour with a high level of habit formation. Car use is an example of a kind of behaviour that is largely based on habits (Verplanken *et al.*, 1994). In fact the formation of habits starts at the moment someone buys a car. Once driving a car frequently, people tend to adapt their attitudes to it. This might well be the explanation of the fact that environmental consequences are so irrelevant to car users, as well as of the underestimated individual financial costs of car driving. Still, there is the question of habit formation. Why are people starting a habit and why has much behaviour a repetitive character?

6.8 Life-styles

Most environmentally relevant behaviour is embedded in on-going behaviour of day to day life. Work, shopping, eating, leisure activities, are all repetitive behaviour and only vary within small ranges. These ranges may vary from person to person, or from household to household, but they show great continuity within a household. Life-style is a sociological concept of daily life patterns, mostly attributed to households. Life-style is determined by two factors: the time budget and the financial budget of a household:

★ How is the income of a household used?
★ How are the persons in the household using their time?

Note that life-style should *not* be considered just another concept of values, general attitudes or behaviour specific attitudes. It referes to patterns of actual behaviour. These pattern may be explained (partly) by attitudes and values, but also by social and physical conditions for behaviour.

6.8.1 *Status and affluence*

Life-style changes have a significant influence on the overall environmental impact of households. The time budget is principally limited to 24 hours a day, the financial budget is still growing in most countries. The obvious effect is that the potential for consumer behaviour is growing, which is probably causing an increase of environmental stress. Every increase in the financial resources of a household will, one way or another, result in new activities, as discussed in section 6.4. Even when someone makes an effort to save energy, this results in a lower energy bill and then the money that is saved will be spent on a new activity. This is a phenomenon called the 'boomerang law of energy saving' (Hayes, 1976), although it is valid for all kinds of consumer spending.

The relation between life-style and consumper spending also becomes manifest in a different way. Dillman *et al.* (1982) showed that rising energy prices resulted in a cut-back in life-style for low income groups. At the same time higher energy prices made higher income groups invest in home energy saving.

The average growth of financial budgets has an impact on status as well. One of the most significant developments in the western world concerning life-styles is enrichment, a trend toward a larger variety of products and services, towards earlier discarding of durable products and more expensive ways of spending leisure time (Uusitalo, 1986; see next section). This is mainly a result of the rising financial budget, but it is also linked to the inflated value of ordinary consumer goods in the positional economy (Hirsch, 1977; also see section 2.5).

6.8.2 Time budget: patterns of day to day life

In principle the time budget is a constant, as every day has only 24 hours. However, there is a tendency towards a slightly larger time budget, because on the average people are living longer and working for a shorter period. Most significant trends however are the changes in time spent for different purposes. The budget spent on work is changing in different ways. Because of emancipation of women the amount of time spent on working has increased over the last decades. However at the same time there is a tendency towards earlier retirement. The division of education, work and personal care within a person's life-cycle and within his weekly schedule has a great impact on day-to-day behaviour and for that reason is determining environmentally relevant behaviour to a large extent.

Although time and money are different budgets, the spending of both is closely linked. In a study on consumption-styles in Finland three main developments were established (Uusitalo, 1986). The first is *modernization*, a trend towards less production within households and an increase in consumption of goods and services. Production in households, which is saving money but also time-intensive, is exchanged for a market orientation. The second development is *enrichment*, mainly resulting from the growth in financial budgets, but with a status component as well (see previous section). The third development is the growing *central position of the private car* in households. The study of Uusitalo established a trend in Finland that is apparent in most industrialized societies, a trend towards a situation already existing in the US. So the environmentally extremely relevant behaviour of car use is very much part of people's life style. To illustrate the links between financial budgets and time spending, we will look again at car use as an example.

6.8.3 Life-style example: the habit of car driving

The most salient attributes for mobility, the time factor and the perceived independence (fig. 6.4) are linked to on-going day-to-day behaviour. These factors, combined with conditions in society like availability of alternative ways of travelling, distances to school or work place etc., determine whether or not people first decide to buy a car and then develop the habit of using it for all purposes. For the decision to use the car, time is the most important consideration. Why is that so important? The answer to this question can be given after an analysis of the time budget of households in different stages of their life-cycle (Batenburg and Knulst, 1993). The study is carried out in the Netherlands, a country in which the dependency on private cars is not yet so dominating as in the US. There are still alternatives such as a fine-meshed public transport system and relatively good conditions for cycling. Nevertheless, car mobility is considered a very serious problem, as the car density (cars per square kilometer) is very high.

Mobility is caused by the need for activities. In table 6.1 the main categories of activities linked to mobility are presented. These are the motives for mobility. Between 1975 and 1990 the number of movements (*per capita*) has increased by 10% in the Netherlands. The type of mobility has also changed. The average number of walks has stayed at the same level, as the typical Dutch category of bicycle transport has increased from 6.9 per person per week to 7.8. The average of movements per week by public transport (1.4) remained constant. Finally, the number of car movements has grown spectacular by 68% to 8.8 per week (6.2 as car driver).

Table 6.1 Motives for mobility as established in a Dutch time-budget study (Batenburg and Knulst, 1993). Average number of movements per week

Activity	Movements	
Education	1.6	
Work	4.2	
Household activities	4.8	(1.0 child care)
Leisure activities	5.7	
Personal care	1.3	
Total movements (1990)	17.5	(1975: 15.7)

There are two reasons for the increase in the number of movements. The financial budget is allowing people more, which is causing a growth in the category of leisure activities. The second reason is that there is a particular part of the population in which most activities (work, household, child care, social participation) are concentrated: the people in the age from 20 to about 50 years old. That part of the population is growing relatively in most western countries. Mobility with the motive of leisure activities is growing fast. However, this is not because everybody is getting more leisure time. Since 1975 people in the age group of 20-50 are spending an increasing amount of time on all their duties together. Hence these people have less leisure time than before. The group over 50 has much less work now and considerably more leisure time. The growing mobility for leisure activities can be found in the latter group. The increase in the number of trips is most apparent among groups that were initially the least mobile. These groups are women, people without a job and people over 65. For leisure activities car ownership is accompanied by a higher frequency of trips. More trips are now also made for reasons of household activities. This growth must be attributed completely to more frequent car use. Car ownership tends to encourage the number of trips: people with a car make more movements for domestic purposes than people without a car. The growth is mainly caused by shopping and carrying children to their schools, clubs etc.

To an increasing extent paid jobs are held by that part of the population that is at the stage of life when not only their profession is time consuming. Other activities, in particular domestic duties such as raising children, and a growing participation in adult education, are concentrated in one part of the population: the 20-50 age group. This leads to the phenomenon of task combination and concentration.

Task combination is the main factor in the growth of car mobility. Within the 20-50 range more people have a job, but domestic duties are spread over more people within this range as well. The combination of tasks demands careful planning of these activities. Most people have a sequence of driving children to school, going to work, picking up their children, shopping, and may be in the evening social participation or educational activities. If anything comes in between the whole planning scheme crashes. That is why these people attribute the greatest weight to time and independence when choosing their way of mobility. Within the age group of 20-50 car use is the standard.

This example shows that, although much of the behaviour is habitual, this does not mean that it is irrational. We have to bear in mind that any habit was first started with an intention that was based on values, beliefs and affects.

6.9 Attempts to change attitudes and behaviour

Originally, as stated in the first section of this chapter, research on values and attitudes started because of high expectations about the relevance of behavioural changes. Anyone who wants to change behaviour through change in attitude, has to go a long way, in particular if the starting point is a communicative message. McGuire (1985, p.259) distinguishes twelve steps in the process of persuasion. Three main stages are *penetration* of communications content, *acceptance* of communications content, and *retention*, the actual adapting of new behaviour consistent with a newly developed attitude, followed by conservation of the new attitude-behaviour pattern. In all stages the process can stagnate, which is illustrated in a simplified way in figure 6.4. The barriers that have to be overcome before a message can result in a behavioural change are linked to the stages in the persuasion process.

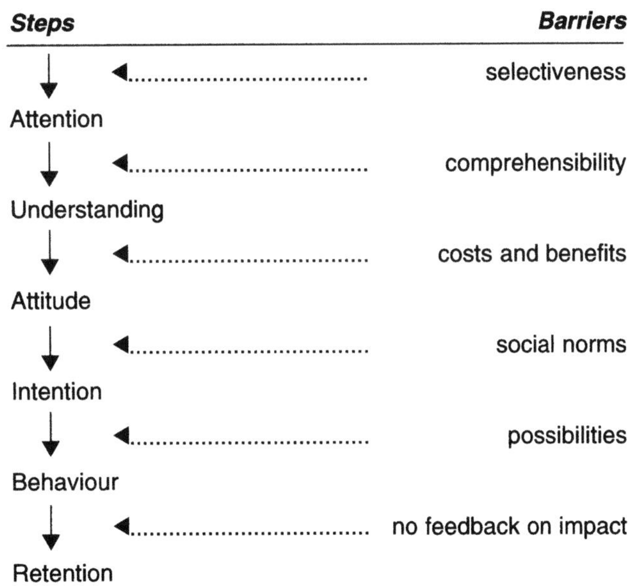

Figure 6.4 Steps and barriers in the process of persuasion directed at behavioural change

We will not describe at length the possibilities for attitude and behaviour change. This can be found in general literature on attitudes and attitude change (McGuire, 1985; Petty and Cacioppo, 1986; Eagly and Chaiken, 1993). Here five general strategies of change are summarized. Some are aimed at changing attitudes first, other try to influence behaviour in a more direct way.

A. Changing incentives.

Changing the subjective utility by introducing new incentives can change behaviour. Incentives are mostly, although not necessarily, financial. It can be a price incentive, for example by introduction of a new tax ('green tax' or 'eco-tax'), or a subsidy for the application of environmentally sound techniques. It can also be financial rewards or penalties, given when some behavioural change has been achieved, such as a certain percentage of energy conservation (Samuelson, 1990). The incentives should be given in the area of salient attributes, so financial considerations should be important in the trade-off. When the primary motives for a particular type of behaviour are not financial, the strategy is not likely to work, it might even be counterproductive, as pointed out in section 6.7.1. Market-oriented solutions for the problem of changing behaviour can be very effective for some types of behaviour, but in other cases they should not be implemented without due consideration.

B. Persuasive communication.

Publicity campaigns are a popular policy option because of their highly voluntary character. Alas, mass media and publicity campaigns are highly ineffective in changing behaviour and even in changing attitudes. First there is selective exposure and then selective attention. People are selective in the messages they expose themselves to and then they often do not notice them. Personal relevance of a subject determines the attention, otherwise the process will already stagnate in the attention phase (Petty and Cacioppo, 1986, p.81). This relevance can either be involvement in the subject or a vested interest in it (Sivacek and Crano, 1982). Next there should be a personal need for cognition combined with the ability to understand the message. According to the TRA an attitude change will only occur when the new information causes a change in the trade-off between salient attributes. However, when it does, such a change is not always in the direction that was originally intended. If there is no personal relevance, for example because the individual has no involvement in environmental issues, changes can only be achieved by so called peripheral cues, such as credibility of the source of the information (Petty and Cacioppo, 1986).

Supposing that the attitude has been changed, all types of barriers (social norms, lack of perceived behavioural control, actual lack of behavioural control) may prevent a change in behaviour. Nevertheless, some studies on environmentally relevant behaviour showed significant results using persuasive information, in particular on a community scale. An example is the study on recycling behaviour of Vining and Ebreo (1992).

C. Information feedback.

More effective in several cases of behaviour with environmental impact is 'information feedback'. It has been successfully applied in cases like household energy consumption (Winnet *et al.*,1978). Information feedback is not aimed directly at attitude change, it is based on learning theory. The information has to be given with a high frequency and the content is not directed at changing the values or the attitudes of the individual. The only purpose of it is to confront someone with the results of his or her recent behaviour. The individual who is receiving the information has to process it himself. He has to link the information with his own knowledge about his recent behaviour. The reason why this strategy is often more effective is that only the individual himself knows exactly his recent past and therefore he is the only one who is able to consider the relevance of information about the consequences of his behaviour. In publicity campaigns only general information is given and this gives everyone an easy way out.

Almost everyone holds the idea that the given information is not applicable to his or her particular personal situation. Hence the effect of behavioural change is probably limited to persons that are already motivated and an attitude change will not be achieved (Samuelson, 1990, p.221).

D. Coercion and mandatory systems of behavioural limits.

Most measures that physically or socially limit the options for behaviour are implemented by law. Many are directed at the level of institutional behaviour, such as emission standards for industry of standards for products. The consumer is buying more energy-efficient equipment nowadays, or cleaner products because manufacturers are producing them, stimulated by government regulations. Measures directed at the consumer level are limited. An example may be speed limits for cars, but these are mostly created for reasons of safety.

There is a wide grey area between changing incentives and coercion. Formal regulations can only be coercive if they are accompanied by sufficient control and effective sanctions. Incentives such as price-raising measures may become behavioural limits as some people can no longer afford particular products (Dillman *et al.*, 1982).

E. Technological changes.

Production can be improved in various ways with regard to various criteria. Cleaner production is a way to limit the environmental impact of a particular product with the result that the same act of a consumer (buying the product) is causing less environmental stress. At the same time products can be improved themselves; for example a refrigerator can be made more energy efficient or CFC's can be replaced by a less damaging cooling fluid. For improving the ways of production and the design of cleaner products, instruments like *substance flow analysis* (SFA) and *life cycle assessments* (LCA) have been developed. In such analysis environmental impact is estimated for stages of the product or for the flow of a substance through society. In an LCA, for example, all impact during the entire product life cycle is assessed, including the consumption stage. Improvement of a product with regard to that particular stage can be achieved by technological means. Technological innovations can be directed at limiting the environmental consequences of the use of a product. Innovations can also be directed at influencing the intensity of use or the way the product is used. An example is a personal computer, that switches itself to state of low energy consumption when the keyboard is not touched for a while, or a truck with a speed limitation device.

6.10 Potential power of attitudes and individual behaviour

In this chapter we have discussed the limitations of the attitude-behaviour approach for solving environmental problems. The unruliness of environmentally relevant behaviour and the limited possibilities for changing behaviour in such a way that it causes less environmental stress, does not imply that individual behaviour will never be the key to environmental conservation. The approach in strategies of changing incentives, persuasion and other policy instruments starts from the implicit assumption that the origin of environmental problems has to be found in the behaviour of individual citizens. Environmental information is basically a top-down activity, based on a *policy belief.* The assumption is that authorities, public bodies and several other agents, such as energy utilities, are well informed about the evil consequences of the ordinary behaviour of the people. The citizens however, are not yet convinced of the true character of their behaviour, so they have to be educated or coached with gentle insistence.

Although environmental awareness at the level of individual citizens is fluctuating, it is generally at a reasonable level. This is not only true for western countries (Kempton *et al.*, 1992), but for many developing countries as well (Dunlap *et al.*, 1993). The point is that actual behaviour stays behind the level of environmental awareness (section 6.6). Pushing back environmental stress by changing behavioural patterns needs structural changes that go beyond the power of individuals. The links between income, consumer spending and environmental stress should be disconnected. The creation of environmentally sound alternatives and pushing back environmental stress per unit of consumption (dematerialization) cannot be achieved by the decisions of individual citizens alone (Von Weizsäcker *et al.*, 1995).

Physical and social limits to changes in behaviour are often more important than attitudes for achieving behavioural changes. As behavioural patterns and life-style are highly determined by the conditions in society and by the absence of important behavioural alternatives, a significant behavioural change can only be achieved within the context of institutional changes and collective decisions. The creation of alternatives for current behaviour mostly requires collective decisions. At the level of values and general environmental attitudes the awareness of the public is much more ahead of policy than authorities think (Kempton *et al.*, 1992). It is where public attitudes have a large impact on collective decision-making were we probably find the potential power of attitudes and individual behaviour. Attitudes and behaviour can be directed at collective and political decision-making. It can be type IV behaviour (section 6.3.2), participation in decision processes, or it can be type IIIb, refraining from particular acts, motivated by environmental concern. In many countries the disappearance of some products, in particular certain types of packaging, is often a result of consumer behaviour. The decisive driving force for the development of less polluting products is demand for such products.

The power of public attitudes is highly dependant on the way decision-making is organized. In decision-making processes many barriers exist that may prevent individuals or organizations influencing the decisions (see chapter 10). An example of this can be found in decisions on waste management and the siting of waste facilities (Wilson, 1996). Although the member states of the European

Union officially have a waste policy acknowledging the supremacy of waste avoidance, minimizing and recycling, in practice the vast majority of waste in all EU member states still goes to either landfill sites or incinerators. In the US however, Wilson states that 'waste reduction in industry is being driven forward largely by community right to know legislation ..., prevention avoids bad publicity.' (Wilson, 1996, p.398). Here successes in waste reduction are attributed to the influence of attitudes of the public formed by information that cannot be withheld.

Victory at Yellowstone

President Clinton's announcement yesterday that a Canadian company has agreed to abandon plans to build a huge gold mine near the border of Yellowstone National Park is a victory for the environment. The proposed mine and its toxic waste would have threatened not only America's first and foremost national park, but also the ecosystem surrounding it. (...) It is finally a belated act of corporate wisdom by the company itself, which ultimately concluded that the substantial profits to be derived from the mine were not worth the certain and lasting public condemnation had the project gone forward....

The New York Times, August 13, 1996.

Another example of the impact of attitudes on institutional decision-making with high relevance for the environment can be found in the story of the dumping of the Brent Spar. Forced by a rapid development in public opinion, triggered by a campaign of Greenpeace, the Shell oil company discarded its plans to dump this large oil drilling platform in the Atlantic Ocean. Authorities in several European countries called on Shell not to sink the platform. An analyses of the events leads to the conclusion that the decisive factor was the attitudes of the public toward the dumping and the oil company (Dickson and McCulloch, 1996). There is a power of public attitudes, but only when the decision-making process is organized in such a way that the public, or organizations, are given opportunities to put the issue on the political agenda. These opportunities do not always exist, as will be discussed in section 9.4.3. In many cases environmental issues are 'organized out of the decision-making process'. In particular public attitudes based on risk perceptions, for instance in decision making on the siting of infrastructure facilities, are often neglected because they do not fit the policy plans of authorities or planners. In chapter 7 we will discuss this kind of perception and the community responses to environmental stress.

Chapter 7

Annoyance and risk

7.1 Impact and reaction

An intervention in the physical environment causes environmental change: a road is constructed, a plant is built, or an airport is enlarged. All these interventions lead to additional activities: new or more traffic, industrial activity, and more airplane take-offs and landings. These may cause any kind of negative feelings, a very broad term that can be interpreted solely as a psychological category, a state of mind caused by lack of personal well-being or caused by any source of psycho-social stress.

In the case of noise, for instance, it is quite obvious that negative feelings may be caused by factors within the physical environment. All the interventions mentioned above produce new or increased noise. The noise may cause negative feelings of all kinds within the communities surrounding the new facilities. In the case of noise psycho-social variables may be affected directly or indirectly by the sounds. That impact may be described in general terms, such as psycho-social 'stress' or 'annoyance'. Besides this psycho-social impact other variables which influence general well-being, like sleeping behaviour or health, may be affected.

In this chapter we will deal with the question of when, and how people perceive human induced environmental impact. Although it may appear self evident that a physical environmental factor like noise has a psycho-social impact, these relationships are not so clear when they are studied in practice. In many cases dose-response relationships appear to be weak. The negative feelings about the exposure to noise could be caused by many other factors than the noise itself. Psycho-social stress may be caused by the daily hassle (family, work, school, etc.), it may be the result of other characteristics of environmental quality (noise from other sources, quality of housing, neighbourhood, greenery, etc.). Last but not least, it may be due to a variety of personal characteristics. In empirical research correlation between environmental changes and negative feelings have been established, but these relationships are always complex.

When reference is made to effects that result from environmental change, the general term is impact. Environmental damage caused by human intervention is called *environmental impact*, while the effects of the damage on society or on individuals are part of the *social impact*. Strictly speaking, social impact is a

neutral term, without any reference to undesirability. Positive effects may occur, in fact the benefits to parts of society are the obvious reason for the human interventions that are causing the impact. When it comes to analyzing the decisions that may cause environmental or social impact, these benefits must be taken into account as well as the negatively perceived impacts.
We are primarily interested in the undesirable effects. As a rule effects from environmental damage are undesirable because one way or another people think some hazardous aspects are linked to them; hazardous for the health of people in general or for human well being. The term risk is sometimes used to refer to all kinds of hazards, but for further analysis we will distinguish here between two major categories: *annoyance* and *risk*. Although the responses to environmental hazards are often described without distinguishing between these two concepts there are some crucial differences. Even within the latter category we can distinguish two types of environmental risks: environmental changes that may occur (probability of causes), and effects on humans that may result from already existing environmental damage (probability of results).

7.2 Annoyance

Disturbances affecting normal day-to-day life may occur in two different ways. Some major physical phenomena cause a constant or frequently repeated social impact. Noise from industry, railroads, airplanes or road traffic is an ever present or frequently recurring disturbance. Other examples are stench caused by manure produced in intensive farming (bio-industry); or dirt and dust caused by industries using coal. Visual impact such as light-flickering by airplanes or windturbines, or new buildings or installations which spoil scenic views may also be considered as a type of annoyance (Wolsink, 1989; also see section 5.2.2).
The psycho-social stress caused by these constant or recurring phenomena may be due to the fact that the annoyance never stops. On the other hand, this may in some cases lead to habituation. The disturbances caused by such chronic factors are called 'environmental annoyance' and the essence of this category is that it always contains a certain amount of sensory pressure (Kastka, 1982), unlike the majority of hazards that take the form of risk. Annoyance may be defined in terms of negative experiences related to physical circumstances within the environment, but in the literature it is also frequently described as a negative attitude towards specific circumstances (Koelega, 1987). We will discuss the issue of impact, annoyance and attitude using the example of sound and the perceptions of noise.

Howard composting site raises big stink in Arundel

It was to be a bridge between two counties to help protect the environment. Instead, the Baltimore Regional Yard Debris Composting Facility is angering nearby residents and prompting a plan to divide the 70-year -old community of Lennox Park in Howard and Anne Arundel counties with a 15-foot wall.

The compost site is on the Howard side of the county line. The residents who have suffered burning eyes, headaches and nausea from its fumes live in Anne Arundel, across a set of train tracks from the facility. (...) 'Lots of times it smells like a garbage truck,' says Joan Fabiszak .... 'Unfortunately they're downwind of the place,' said Howard County Executive Charles I. Ecker (R) of Arundel residents like Fabiszak, whose yard face 400-foot-long steaming rows of rotting leaves and shredded Christmas tree.

The Washington Post, February 22, 1996

7.2.1 Sound and noise

Sound is a physical phenomenon. Mechanical interventions in the environment cause a wave motion in the air, or in some other medium, which in turn causes a physiological response in the human ear. Then we 'hear' the environmental change and the sound becomes a sensory perception. This is a mental event which immediately starts a process of interpretation. Sounds may be enjoyed, they may be perceived merely as functional, or in some cases they are felt as unpleasant. When sound is perceived as noise, we are usually referring to sounds which may be classified as unwanted.

Noise is defined in practice as audible acoustic energy that adversely affects, or may affect, the physiological or psychological well-being of people (Berglund and Lindvall, 1995, p.4). These effects may be very serious, such as, for example, permanent hearing loss caused by frequent or long exposure to high levels of sound. However, this loudness is often due to sources with partly voluntary exposure, so the noise cannot be qualified as an environmental problem. These effects are mostly found among workers in noisy industries and among people with specific leisure time activities such as listening to music through headphones (Axelsson, 1991). Yet another far reaching effect may be the impact of noise on the cognitive abilities of children. Disturbances of cognitive development, like reading deficits, are more frequently found among children who are frequently exposed to high level noise (Wachs and Gruen, 1982). Many other possible effects of noise on humans are reported: aural pain, interference in communication, sleep disturbances, mental disorders, detoriation of task performances like reading or working, decreased social behaviour (reduced helpfulness, increased aggression), and even direct physiological effects like elevation of blood pressure (Berglund and Lindvall, 1995).

When we are interested in the impact of noise on the well being of people, measurement of the physical loudness of the sounds is far from sufficient. In practice, physically measured sound pressure levels are weighted for sound characteristics in accordance with the response character of the human auditory system. Several standard 'weighting filters' exist for the measurement of sound pressure and the most commonly used is the so called A-weighting, a correction on the physical sound pressure for the frequencies of the tones.

7.2.2 *Subjective perception of noise*

Calculated sound pressure levels, even when they are corrected for human perception, are by no means adequate indicators for the effects of noise on humans. The perception of sounds is a highly complicated matter, which is only partly determined by the sound energy. As well as the sound energy (frequency and amplitude) there are several other factors determining the way people perceive environmental sounds: the duration of the sounds, the time of day, and the impulse character. Even when we take these factors into consideration, we still do not know whether or not sounds are interpreted negatively and qualify as 'noise'.

Indicators for general noise annoyance in a research questionnaire
(example of items used)

For each of the situations I read out, would you please use this opinion thermometer to give a rating between 0 and 10 of how much annoyance you feel?
(Record: 0 - 10 rating)

Firstly, how much annoyance do you feel
- when a pneumatic drill or jackhammer is operating nearby
- when you are woken up by a dog barking
- when you hear military gunfire or explosions
- when people are talking when you are watching TV
- when someone rustles paper at the movies
- when you are trying to concentrate in noisy surroundings
- etc.

Source: study on artillery noise by the University of Sydney, Australia (R.F.S. Job, personal communication).

Several indicators for reaction to noise exist (Kryter, 1985). These are indicators based on questionnaires with item-clusters. In surveys items can be used to build specific indexes, such as for example an index of the degree an individual is

generally affected or not by noise. Other indexes are constructed for the measurement of annoyance, or as an indicator for disturbance of daily life.
The concept of annoyance is commonly used to describe the general impact of noise on communities. The dose-response relation between noise exposure and average annoyance rates is fairly high (Schultz, 1978). Figure 7.1 shows the percentage of the population that will be greatly annoyed as a function of noise exposure. The curve was originally based on 161 averages derived from surveys of community reaction to transportation (traffic, railways, aircraft) noise. The curve has been validated with many other surveys ever since, and the results for sources of noise other than transportation, for example power stations, appear to be similar to the synthesis curve of Schulz (Job, 1989).

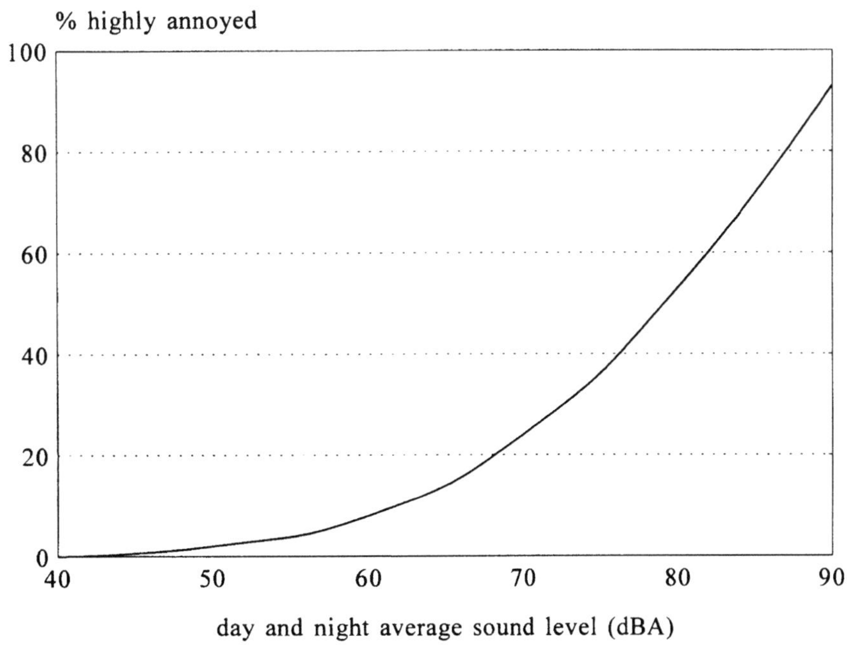

Figure 7.1 Percentage highly annoyed and day-night average sound levels (dBA) for transportation noise (source: Schultz, 1978).

Although average levels of annoyance can be predicted fairly well, the dose response correlations at the individual level are generally rather weak. Psycho-social factors like personal sensitivity and attitudes are due to a large variation between the annoyance. In fact these variables are better predictors of reaction to noise than noise exposure (Job, 1988). In particular, attitude to the source (the degree to which someone favours or disapproves of the source of the noise; also see section 6.6.1), is highly correlated with reaction. When someone dislikes the source he will more easily feel annoyed by sounds emitted by that source, although the reason for disliking the source may have nothing to do with noise. Interpersonal variation is hardly explained by differences in noise exposure, but largely by personal factors and the social context. Actually noise annoyance itself is probably more an expression of attitude to the source than a response to actual sound pressure, so there is high level of subjectivity in annoyance. It should be emphasized, however, that the conclusion by no means might be drawn, that actual sound pressure levels are not a significant environmental disturbance factor. The dose-response relation as shown in figure 7.1 is fairly strong, looking at average figures of community reactions.

7.3 Risks: eventual causes or eventual effects

At first sight risk may seem to be a category of phenomena that can be defined objectively. An environmental risk is the probability that a certain objectively defined negative effect caused by other parties will occur, due to environmental factors. Risks that people take themselves, such as accidents on skiing holidays or lung-cancer by smoking cigarettes, are not considered as environmental risks. As we will discuss later (section 7.5.2) the fact that environmental risks are not taken voluntarily, is particularly significant. There are two types of environmental risks: events (causes) that may occur with a probability of less than 100% and effects on humans that may result (also with a probability less than 100%) from given environmental damage.

The *risk of undesirable events*, the risk of 'potential' accidents, belongs to the first type. Many kinds of environmental impact do not have a chronic or frequently recurring character. In fact many of them do not actually occur at all: there is only the possibility of an event that will cause environmental impact. As long as these events do not occur, these incidental effects do not have physical consequences. Nevertheless, incidents may happen and so they represent environmental hazards that should be taken into consideration. It is the type of risk insurance companies are dealing with, although insurance charges will rise as a result of environmental damage.

The second type involves a number of already existing examples of environmental damage, sometimes as a result of catastrophes in the past. The environmental damage seldomly affects humans directly. Contaminated soil can cause health problems, but in many cases it does not. As a matter of fact there are few cases in which scientifically reliable cause-effect relations are demonstrated. When toxic substances are found in a residential area and there are actually people suffering from real health problems, as a rule it will only be a small part of the population. Environmental damage represents a *health risk* here: there is no guarantee for anyone that he will be among those unaffected by such pollution, but at the same time nobody knows whether or not he will be affected. Although the probability of a real effect is mostly quite small, the risk is present for all.

Global warming hits house insurance
Costs quadruple as weather damage worsens

Homeowners face crippling rises in insurance charges - with companies blaming global warming for the increase. Insurance costs have quadrupled over the past 30 years because of weather-related damage. And as the climate changes further, insurance costs could double again in the worst affected areas. The government has been told by leading insurers that changes in the climate - caused, scientists believe, by pollution - will increase damage to house from draught, storms and flooding. Most at risk are householders in southern England and the Midlands, whose homes are built on clay. (...) Dr. Dlugolecki said 'Insurance companies studying the statistics have realised that droughts are particular problems for homes built on clay.. So insurance premiums and costs in those areas have increased far more. We believe this is related to climate change, and we cannot imagine the situation reversing. Droughts are becoming more frequent and more nasty.'

The Daily Telegraph, October 14, 1996.

7.4 Risk: a subjective concept

Because the concept of probability could be considered merely 'statistical', a number of studies have been carried out to establish whether or not 'risk' really is an objective concept. As in the case of annoyance, neither are risk clearly defined objective phenomena. Within the scope of risk-analysis and risk management three main streams exist, that will be discussed in several sections in this chapter: a policy and culturally oriented approach, a technological approach and a social-psychological approach.

In risk studies the specific definition of risk is always the key issue. For any attempt to measure or assess the amount of risk, the definition is determining the outcome. Femers and Jungermann (1992) classified all different types of definitions used in risk studies in five categories:

- Measures for objects under threat (numbers, amounts, time-measures, spatial measures);
- Measures for causes and events (accidents, disturbances, death causes);
- Measures for probabilities of damaging events;
- Measures for consequences of damaging events;
- Measures for combined probability and consequence.

What type of definition is used is linked to the approach. In the following sections we will discuss the three parts of the risk-concept: probability, the seriousness of consequences and the range of consequences. In all three components quantification and qualification are dependant on basic assumptions about the way the risk should be approached. In particular the amount of risk linked to the use of certain technologies has a strong cultural and ideological component as we shall discuss in section 7.8.

7.4.1 The concept of probability

At first sight the probability that an event occurs may be viewed as an objective datum. However, probability implies uncertainty, and uncertainty may originate from different sources.

– *Statistical uncertainty.*

From general knowledge about the probability that some event may occur, no certainty can be derived for individual cases. While the percentage of breathing problems as a result of photochemical smog is known (and not controversial) it is still unknown who will be a victim and who will not. Although some knowledge may be available, for example, based on knowledge about well defined groups with an increased risk, there will always be statistical uncertainty regarding individuals within that group.

– *Lack of knowledge.*

Uncertainty about the probability of events may be a result of the fact that there is no sufficient knowledge available. Often it is simply not known which environmental impact has to be linked to certain activities. For example, the enhanced greenhouse effect has been known for more than a century. However, because of complex relationships between all components of the atmosphere and the factors influencing them, the future environmental impact of the enhanced greenhouse effect is still unknown (section 10.5). Large research efforts will be required to produce the knowledge needed.

In some cases of enormous environmental long term risks, sometimes on a global scale, there is not only little knowledge, there is even no perpective available to achieve the level of knowledge needed for a significant reduction of uncertainty. On the subject of decreasing global biodiversity for example (section 10.6), we are certainly dealing with a significant long term risk. While 'knowledge of the identity and relationships of disease organisms and their vectors is required to counter present and future health threats', the current infrastructure in the scientific field of systematic biology appears to be totally insufficient for reaching the necessary level of knowledge (Blackmore, 1996, p.64).

– *Lack of agreement.*

In many cases there is scientific knowledge available, but there is still uncertainty caused by scientific disagreement. Knowledge about the processes that lead from causes to effects are not sufficient for an assessment of risks. Basic assumptions appear to be significant in scientific discourses most of the time. The global change due to the enhanced greenhouse effect is again a good example. There is not only a lack of knowledge, there is also a wide range of scientific opinion about this phenomenon. As a result the social consequences of it are also highly uncertain.

Even in cases in which probabilities are reduced to statistical uncertainty, calculated risks can be viewed differently. In the current technocratic approach the probability of an event is defined in terms of relative frequency. However, a different approach is possible, based on 'Baysian' statistics. The crucial difference is that currently probabilities are estimated starting from the assumption that nothing goes wrong under 'normal' conditions (the 'null hypothesis'). In a Baysian approach the probability of an undesired event is not assumed to be zero, it is *a priori* set at a reasoned probability. The fact that the starting point under 'normal' conditions is a probability larger than zero may be much more realistic, but the '*a priori* assumption' about the actual probability rate is a matter of opinion. For instance, an opinion about what exactly may be considered 'normal' conditions. As a matter of fact the subjective Baysian approach to probabilities recognises that uncertainty is partly a result of different views. So calculations are based on another 'null hypothesis', a value which can be adjusted for changed conditions or growing knowledge.

7.4.2 Seriousness of consequences

The second component of the concept of risk is strongly determined even more by subjective factors than the concept of probability. It starts with the positive or negative valuation of environmental change. According to the definition of environmental risks we are only interested in undesired effects. Not all environmental changes are considered to be negative by everybody.

The valuation of environmental change and its social consequences in particular are culturally determined. In an environment which is already a result of human intervention, changes are sometimes considered to be improvements. This particularly holds true for the built environment, in which facilities are factors in the quality of the living environs as much as ecological factors are. In particular the esthetics of intervention are subject to changing valuation. Although areas that are relatively undisturbed ('wilderness') are highly valued nowadays, it has not always been the case. Thomas (1990) described an interesting turn in thinking about lands that are not yet cultivated. 'Wastelands' used to be described as useless and regions with lots of wasteland were seen as backward. Nowadays, cultivation of scenic areas is considered undesirable, the valuation of nature has increased and many nature reserves have been established (section 4.4). Accordingly potential damage to these undisturbed areas are now regarded as risks.

For most environmental impact the negative valuation is not in dispute. However, the seriousness of the effects is a matter of controversy. For risk assessments and environmental impact reports attempts have been made to classify the consequences of environmental impact. The valuation of these consequences fits in with the general theory of human motivation by Abraham Maslow. He distinguished, in this psychological theory of human needs, a hierarchy of drives: four levels of motivation based on deficiencies and one level of motivation based on possibilities for growth (Maslow, 1943; 1971). These five levels are, in order of fulfilment:

1. physiological needs (basic needs like food, drink, sleep)
2. safety (personal needs)
3. love (social needs)
4. esteem (respect)
5. self-actualisation (growth motivation)

Maslow states that the hierarchy is partly based on instinct. The lower the ranking of needs, the more significant the fulfilment of these needs will be. Excessive frustration of the deficiency motives (1 to 4) will cause psychological, or even physiological, illness. As causes for frustration we should not only consider matters of life and death here, but situations that might be threatening as well. Dangers that originate in our environment and in particular hazards that are linked to environmental damage, will put pressure on the fulfilment of human needs. The valuation of the fulfilment of needs of a higher order will be low when the needs at a lower level are not fulfilled. This implies, for example, that a person's valuation of social relations (level 3) decreases when air traffic noise frequently disturbs his sleep (level 1) or when his feelings of safety are seriously affected by perceived risks of air traffic accidents (level 2). In practice, some evidence has been found that 'feelings to be left alone' are stronger indeed among people who suffer from frequent sleep disturbances caused by high levels of traffic or aircraft noise (Öhrström, 1989). The same would hold for a severe risk, for example when someone has strong feelings of insecurity because of a facility nearby that is considered dangerous, or because of the existence of a refuse tip containing hazardous toxic substances. According to the theory of Maslow the significance of a disturbance is not limited to the need affected itself, the value attributed to higher ranked needs decreases as well, although they may not be affected directly.

The hierarchy of needs is not absolute in such a way that all needs of higher ranking have be fulfilled before the next become important. Nor it is easy to indicate to which level needs should be attributed (Wahba and Bridwell, 1976). Nevertheless Maslow's theory has appeared to be fruitful and the seriousness of the social consequences of environmental impact can be derived from the hierarchy.

7.4.3 Range of consequences

The significance of many environmental problems will primarily be derived from risks for individuals. The possibility that in cases of environmental damage many individuals are at risk simultaneously gives environmental risks an extra weight.

Individuals may suffer from noise exposure, for example, with sometimes very severe *personal* consequences for their health (section 7.2.1). However, for society it only becomes an issue when certain critical numbers are exceeded.

Sometimes it is not just a number of individuals, but a whole *community* which is at risk, particularly in cases of locally concentrated facilities with a risk of disasters, like nuclear facilities or particular chemical plants. This adds a new dimension to the public responses to perceived risks (see section 7.5.2 and 9.8). The community at risk can even be larger than a local community. In chapter 10 we will discuss the phenomenon of globalization, a trend which is partly based on growing awareness of environmental threats of global scale. Some environmental issues are defined in terms of global change and in terms of *global risks*. These issues are about world-wide hazards, such as climate change, that may threaten large communities or even whole nations. However, as Yearley (1996) clearly demonstrates, it has also become a strategy to increase public and political attention for environmental issues when pollution, resource depletion, loss of biodiversity and other types of environmental hazards, are defined as global risks.

Yet another dimension, of the range of consequences is the time horizon. In section 3.4 this issue was described, and in section 5.2.2 the transfer over time as a basis for social dilemma's was discussed. The benefits of most activities will be for the present generation, as much of the environmental impact becomes manifest for future generations. How far in the future will the consequences of environmental change become manifest? In general the impact is perceived as less significant when the time horizon is far away. Weighing benefits against environmental risks is in practice a matter of cost-benefit analysis. In economic analysis a general assumption is that a unit of cost or benefit matters more when it is experienced here and now than when it is going to occur in the future. This decrease of importance of future gains and losses is known as *discounting*, which is common practise in cost-benefit analysis. Sums of money in the future are computed to their present values and in these computations discount rates are used. The justification for discount rates lies primarily in the existence of interest rates, but uncertainty about the consequences adds to it and uncertainty is part of the concept of risk. Uncertainty is expected to increase with time and hence it can also be expressed in the form of a discount rate (Pearce and Turner, 1990, p.218).

The discount rate is an annual correction factor, which is slightly less than one. It causes a small present weight of consequences that will be manifest in the long run. Beside the problems that exist anyway with the computation of external costs, discounting results in a very low present value for future costs. Cost-benefit analysis of future risks, like the risk of global warming as a result of the enhanced greenhouse effect, will easily lead to the conclusion that measures are not expedient now.

The discounting practise in risk analysis is criticized from an approach which emphasizes the requirements of sustainability (Pearce and Turner, 1990, p.225). According to the WCED (1987) sustainability is the fulfilment of present needs without compromising the ability of future generations to meet their own needs. It would mean that present values of future environmental risks should not be discounted to a level at which they cease to affect present decisions. Some risks, in particular those showing little cost in the near future, would become serious cases should discounting be limited in the light of sustainability. Examples are the concentrations of hazardous substances in oceans, such as PCB's, all activities that may result in ground water pollution (which are many), or long term exposure to substances that affect genetic material.

7.5 Perceptions of risk and annoyance

As already noted all three components of risk, the probability, the seriousness of consequences and the range of consequences, have subjective meaning. No matter if we are dealing with expert- or lay-judgments, values partly determine the perceptions of risks and annoyance.

7.5.1 Undesirable events and their probability

If people make estimations of probabilities in advance, these appear to diverge strongly from actually established figures. Some causes of this systematic bias in the attribution of the probability of events are known. Since the early seventies a research tradition has developed in cognitive psychology directed at the ways people come to judgments in situations of uncertainty (Kahneman *et al.*, 1982). A large number of studies concern catastrophes of any kind and environmental risks, but many others concern quite different events, sometimes of a trivial nature. The results of these studies are often applicable to situations in which people have to make judgements about possible environmental changes or about environmental hazards.

Kahneman *et al.* (1982) distinguished several *heuristics* which the human mind follows in the process of judging probabilities. In controlled laboratory conditions these heuristics are demonstrated. However, in real life they are not so easily recognised as all forms of heuristics are interwoven. Some features of judgment in situations of uncertainty are evident, for example the frequency of particular events is systematically underestimated as the frequency of other events is overestimated. As a rule, small probabilities are overestimated, as high probabilities are generally underestimated. Some of the heuristics are particularly interesting in relation to environmental risks.

The first heuristic and origin of bias in probability assessment is *representativeness* (Tversky and Kahneman, 1982, p.85). Suppose in a small town two different plants are located. One is the source of an emission that is causing stench, while the other does not cause such annoyance. This fact is common knowledge, because papers and broadcasting stations frequently report the problems that arise from these emissions. When people not living closely to this town are asked which plant they expect to produce more noise annoyance, the stench producing plant will on average score higher. For this plant stench will be viewed as representing annoyance and as a result other types of annoyance will more easily be associated with this plant.

The second probability heuristic that is relevant for the assessment of risk is *(cognitive) availability* of information that is strongly associated with the risk. In perceptions of environmental risk this is probably the most important factor causing bias in estimates of probability. The expectations about the frequency of events is derived from the cognitive availability of information recalled about similar cases (Tversky and Kahneman, 1973, p.209). Summarized, it would mean that someone will assess the probability of toxic substances in the soil of his garden relatively high when a case of toxic soil pollution has recently occurred anywhere in his neighbourhood, for example in his own town or in the community of a relative or good friend.

A third source of bias is *anchoring*. People show a tendency to adapting their judgements about probabilities of events to values that are given as a starting point. The probability becomes a derivation from some other anchor-value which is previously known. Suppose in the past year the air quality in a particular area on five occasions reached a level which caused the authorities to call on asthmatic patients to stay at home. When people are asked to estimate the probability that such events will occur twenty times or more in the next year, the anchor value is five. As a result there will be a tendency to underestimate that probability. In an example (in frame) the average estimates of two groups with different anchors are given. The high figures are produced by a group with a given anchor of a mortality rate of 50,000 for motorized mobility, the low figures came from a group were given the anchor of 1000 for electrocution. The estimates will in general be closer to the anchor values than the expected frequencies actually are, so a low achor value causes low expectations whereas a high anchor value results in relatively high expected frequencies.

Some examples of average estimated frequencies of death causes in the US

Cause	Actual frequency	1st Anchor Group	2nd Anchor Group
Fireworks	6	331	77
Lightning	107	128	64
Electrocution	1025	586	1000 *
Traffic accident	55350	50000 *	33884

source: Lichtenstein *et al.*, 1978.

The assessment of the probability of events is biased by interactive effects between cognitive processes and the social context of the events. Apart from this cognitive component of risks, the evaluative component in judging actual impacts of the potential range and seriousness of consequences are important factors in the perception of risks.

7.5.2 Catastrophality

The systematic biases in judging probabilities described previously are the reason for some risk analysts to declare public perceptions of risks as irrational. For experts in assessment of probabilities of accidents with nuclear plants, for instance, a real accident is no falsification of their calculated risks. The probability will always be more than zero and an expected frequency of once in 500 years means an accident may happen next year as well as after 300 years. If in the case of an accident a particular good reason could explain an increased probability, for example mismanagement in the case of the nuclear power plant near Chernobyl, that accident fits perfectly within the risk perceptions of the experts.

The public however, will react differently. Although attitudes 'are very resistant to fundamental change', the attitudes on nuclear plants and their risks changed as a result of the Chernobyl disaster (Eiser *et al.*, 1995, p.159). In a study in eleven European countries a sharp change took place immediately after the accident. The effect was moderated later but it did not disappear (Renn, 1990). Chernobyl was a far-reaching event that has also stimulated nuclear experts to re-analyse the working of existing reactors. The radical shift in risk perceptions can easily be explained from the heuristics described in the previous section. Because of the enormous exposure to information about the accident in the mass media, overestimation of the probability of nuclear accidents could easily arise because of the probability heuristic of (cognitive) availability (previous section). However, in the meantime, the shift in risk perceptions and attitudes to nuclear power appear to have changed more fundamentally. The shock-effect is now over, but the resistance towards nuclear power in many countries remains at a high level long after the Chernobyl accident.

Many surveys about the risks of nuclear power, long before Chernobyl, have shown that supporters of nuclear power base their judgments on totally different aspects from those chosen by opponents (Eiser and Van der Pligt, 1979; Woo and Castore, 1980). For one group the probability of accidents and the consequences for environmental impact and health, are relevant. For individuals who base their judgment upon these risks other aspects like economic significance are irrelevant, no matter what the assessed likelihood of these aspects are. For others the attitude is primarily determined by their judgment of the economic benefits of the application of nuclear energy. They show a tendency to ignore the issues of safety and health-risks. The fundamental change in attitudes to nuclear power is not only a result of changes in probability estimations, there has been a shift in the importance that people attribute to the costs and benefits of nuclear power. The risks of nuclear accidents are now relevant for a larger part of the public than they were before. Many people who based their opinion on the economic significance of nuclear energy started to weight the probability of accidents, whereas they had previously ignored them. As a result of the catastrophe the severe impact of such an accident has become clear to many and it appears that this has changed the assessment of the risks people attribute to nuclear power. To sum up, if the range and seriousness of consequences are enormous, very small probability rates become fairly irrelevant. Slovic *et al.* (1985, 1987) established a clear pattern in the assessment of risks. With multi-dimensional scaling methods they mapped the judgments about risks from many different sources. The result was a pattern that re-appeared when mapping the result for very different groups. This fundamental pattern underlying risk-perceptions contains two qualitative judgments, both uniting several analytical distinctive properties of risks (Slovic *et al.*, 1985, p.103). Although striking cultural differences for some particular risky activities were established in a comparison with a Japanese sample, the two dimensions in the cognitive respresentation appeared similar in the American and Japanese cultural setting (Kleinhesselink and Rosa, 1991, p25).

I – The first factor is *'unknown risk'*. In this judgment factor the degree to which the risk is scientifically known, the degree to which the risk is known by the exposed, and how familiar people are with the risk are expressed. Within this factor 'unknown risk', the degree of voluntariness and the degree to which the risk can be controlled, is also expressed. These factors may be analytically distinguished, in cognitive patterns, however, control and voluntariness are apparently strongly associated with familiarity and knowledge.

In the cognitive process of making judgments people strongly associate environmental risks with non-voluntariness and at the same time with a low degree of familiarity. Relatively unknown and non-voluntary risks, for example, are the ones associated with food colouring and preservatives, and with pesticides and nuclear power. Because of the cognitive association of nuclear power with all other nuclear activities, including nuclear weapons, in the Japanese sample nuclear power was judged relatively 'known', whereas in the US it was 'unknown'. Generally considered well known and voluntary are risks linked to mountain climbing, swimming, skiing and other non-environmental risks, which are not environmental risks by definition.

II – The second qualitative factor of judgment is about the range and seriousness of risks. This *'dread risk'* factor is the expression of dread and the certainty that the event will be fatal. At the one hand with a relatively low degree of catastrophality we find risks attributed to lawn mowers, home appliances, football and food colouring. On the other hand we find catastrophic risks associated with aviation, hand-guns and nuclear power.

Basically similar patterns were found for groups with very different composition: women voters, students, political activists, and for a group of risk-experts as well. In some cases the assessed probabilities of risks may diverge widely when lay-judgments and expert-judgments are compared (section 7.6.3), but the aspects that appear to be significant are the same for both groups. This indicates that there is no reason to call lay judgments irrational and expert-judgment pre-eminently rational (Fischhoff *et al.*, 1987, p.1114). It is another kind of rationality, since there are sometimes large differences in the probability assessments themselves. An obvious example is the judgment of the risks of nuclear power. Laymen and experts agree on the nature of the risks: they are unfamiliar and non-voluntary, long term and with far reaching impact. The differences can be found in the actual estimations of the probabilities of accidents and in the decisions that are based on the combination of the qualitative judgment and the assessed probability. The point is that people who feel threatened, are 'unlikely to be more than partly reassured by a few statistics. We do not tend to think of danger in a formal statistical manner, so an (exaggerated) perception of danger cannot simply be attributed to a misperception of statistics.' (Eiser *et al.*, 1995, p.180).

7.5.3 Decreases of perceived risk

If the judgments of residents living near nuclear power stations are compared with the judgments of others, many neighbours seem to have become familiar with the facility (Eiser *et al.* 1995). In the course of time opinions tend to become more favourable towards the application of nuclear power. Probabilities of negative impact in the long term are viewed more optimistically by people with a reactor in their neigbourhood (Van der Pligt *et al.*, 1986 p.89). The same holds good for people who work in the nuclear industry.

Explanations of these differences can be found in behaviour as well as in changed attitudes. People who do not accept the risks can move out of the region and they will be replaced by others who are less rejective. The average acceptance level rises this way, although the psychological stress for the ones who cannot move, for example because they cannot afford it financially, may be worse as they feel themselves 'captives' like residents in a polluted residential district (section 7.6.3). Another explanation for decreased levels of risk perception is *dissonance-reduction*. This phenomenon is part of the of the theory of cognitive dissonance (Festinger, 1957). When a person holds beliefs and views which do not correspond with his behaviour and his perception of the situation, this person is in a state of dissonance between his cognitions, which is causing feelings of discomfort. To reduce the stress-feelings the cognitive dissonance should be reduced, which can be done by two strategies. Anyone perceiving his situation risky because he is living near a nuclear reactor, could either move out (change of behaviour) or repress his risk perceptions. Ester *et al.* (1983, p.360) suggested

that people living near a nuclear power station are repressing significance of the negative consequences, motivated by a lack of cognitive consonance. These processes of dissonance reduction can be significant in many cases of risk and annoyance (Shippee *et al.*, 1980).
Familiarity with an environmental risk may arise as long as the hazard is diffuse and not acute. In most cases of pollution or disasters there are no immediate casualties. The psycho-social impact is limited to stress caused by annoyance and threat. As long as this stress stays at a level at which people can cope with it by either adapting their behaviour in a simple manner (insulation of their house against noise, or closing windows), or by coping with it psychologically (dissonance reduction), then it is possible that people build up familiarity and get used to the environmental impact. When the stress becomes too great, however, psychological strategies for coping are not sufficient any more. In cases of severe pressure caused by obvious environmental damage with high levels of perceived risk, new mechanisms of behaviour change become significant (section 7.6.3).

7.6 Risks and decisions

The responses of society following the recognition of environmental hazards may vary widely, depending on all three components of risk, the probability, the seriousness of consequences and the range of consequences. However, judging environmental impact and its hazards in terms of relative acceptability does not make sense. The basic idea is that talking about risks and hazards always means talking about decisions that society has to take (Fischhoff *et al.*, 1987). According to Beck (1992) the decisions about risks are even the most fundamentally a post industrialized society has to take, a notion on which he based his concept of 'risk society'.

7.6.1 Acceptable options and decisions about risks

All discussions about the acceptability of risks or about negative effects in general, cannot be meaningful unless they are held in a decisional setting. Any option can only be judged as acceptable or not acceptable if it is compared with other options. As an option may be acceptable in comparison with other options, the term 'acceptable risk' is not appropriate for describing the results of a decision-making process. The fact that society has decided to use certain technologies, or to build certain facilities, does not mean that the risks involved are acceptable. 'The risk associated with the most acceptable option will by no means be acceptable in an absolute sense. One accepts options, not risks, which are only a feature of options. Moreover, even the choice of an option is highly contingent on how a problem is defined, on the other available options and who is doing the deciding.' (Fischhoff cs, 1981, p.139).

All these factors make risk into a phenomenon which cannot be rated absolutely and objectively. The decision which option of a set is the most preferable, is inherently situation specific. This becomes clear particularly when we look at perceptions of risk at the level of individuals and the community that has to decide about taking new risks or dealing with old risks (section 7.6.2). Risks are social and cultural constructs of which the valuation is time-, place- and situation-dependant (Covello and Johnson, 1987).
Since there are no universally acceptable options, risks and other costs or benefits are even less universal. Although there is no universal concept of acceptability, patterns in perceptions of risk are nevertheless related to general values. Moreover some values, in particular general environmental values (section 6.5) may be affected by experiences with risks and decisions about risks in the past. In an overview of the development of environmental awareness, Hays (1987) concludes that environmental values are strongly connected to the risks people perceive in their own life. Environmental values are significantly infused with a sense of place. In thinking about environmental quality people tend to link it to their own place, where they live, work, recreate and so on. This tendency to defend the community and the home triggers an overall awareness of environmental quality (Hays, 1987; also see section 9.8).

7.6.2 Community response: facility siting

Because all risk judgment takes place within a decisional setting, the moment at which an intervention with environmental impact is proposed is particularly significant in the development of judgments about risk. The construction of facilities such as roads, railways, airports, power plants, waste incinerators or nuclear repositories etc., requires land and it affects the environment. The process of siting facilities is a manifold source of local political conflicts in which risk and annoyance are vital issues. Deliberations on the best location may go on for years, while the start of construction is often delayed owing to public opposition to the project. The term *'LULUS'* ('locally unwanted land uses') is frequently associated with these new facilities (Popper, 1985). Another term that often crops up to explain the phenomenon of harsh resistance is *'NIMBY'*. It stands for 'not in my back yard', an attitude which is thought to be behind much local opposition. The term 'nimby-syndrome' has a highly negative connotation. Cynical motives are attributed to those who 'suffer from' this syndrome and opponents of siting are described as 'short-sighted'. In a study on community response on nuclear waste siting it is stated: 'in the typical siting case, much of the opposition is parochial and short-sighted, as captured by the term nimby.' (Easterling, 1992, p.469). The implication is clear: this selfish opposition is merely based on self-interest. Proponents of this argument do not distinguish between the interests of the opponents and their motives and they tend to disregard the risk perceptions of the opponents. In the literature on facility siting and decision processes this not-in-my-backyard 'explanation' of local resistance has become outdated. However, in policy it is often still considered common knowledge (Freudenburg and Pastor, 1992; Wolsink, 1994).

The nimby-effect or -syndrome is commonly put forward as a ground for opposition, especially by bodies in favour of the construction of installations. A strong example can be found among the proponents of nuclear energy. As a rule, however, opposition to nuclear installations is connected with the fact that many people are against the use of nuclear energy. They are opposed to nuclear facilities in anyone's backyard, including their own. In the US the shift in the public opinion from initial (majority) support to (majority) opposition against construction of local nuclear plants took place between 1980 and 1982. This is correlated with the shift from support to opposition against building new nuclear plants anywhere in the US (Rosa and Freudenburg, 1993, p.48 and p.57).

Opposition can take four forms, only one of which fits the definition of the nimby syndrome (Wolsink, 1994, p.862). The following characterization is based on a proposed gene technology plant, but it is equally valid for other facilities.

Type A: A positive attitude towards genetic modification, combined with a rejection of, and opposition to the construction of, a genetic modification plant anywhere in one's own neighbourhood. This type of perception-behaviour combination reflects the only true nimby stand-point.

Type B: Rejection of, and opposition to a genetic modification plant in the neighbourhood, because one is against the technology of genetic manipulation itself. This position is in the literature sometimes called 'niaby', not-in-any-backyard.

Type C: A positive attitude towards genetic modification, which turns into a negative attitude as a result of the discussion surrounding the proposed construction of a plant in which genetic modification is applied. This is the dynamic niaby variation, which reflects the development of risk perceptions during the decision-making process.

Type D: Resistance created by the fact that some projects are themselves considered faulty, without a rejection of the technology itself. In the case of genetic modification it is possible that people are not opposed to genetic technology under all circumstances, but do not accept the conditions they are confronted with, for example when they are not convinced the control is adequately organized. They might accept some applications, for example for medical development. At the same time this does not mean one is accepting the technology for other applications, for example, in genetic modified animals or for improved resistance against herbicides in modified potatos or soybeans.

All four types of motives with different risk perceptions will exist in any case of facility siting. Beside the real self interest motivated nimby-part of the public (type A), nuclear power plants are the textbook example of the not-in-any-backyard type B. However, many people shifted in the past from support to opposition during a public discussion about the siting of new nuclear facilities, which is an example of type C: developing opposition (Blowers *et al.*, 1991). Type D has been recognised among supporters of the application of nuclear power as well (Eiser *et al.*, 1995, p.201).

The risks people associate with the facilities are the main reasons for opposition. Risk judgments become salient factors as a result of the decisional context. Decision making on facility siting is putting environmental risk on the political agenda (see section 9.4.3). In these political processes the differences between the assessments of risks by experts and the social rationality of public perceptions become apparent. In the communication the typical pattern is that the public is asking 'questions that are not answered by risk technicians at all, the technicians are answering questions which miss the point of what is really asked and what feeds public fears.' (Beck, 1992, p.30). It is the same pattern we can recognize in disputes about the management of risks that resulted from failures in environmental management in the past.

7.6.3 Community response: manifest environmental impact

The first time soil pollution was recognized as a real threat to the health of humans was in the case of Love Canal (Levine, 1982). In a district of Niagara Falls, NY, the residents of about a thousand newly built houses were confronted in August 1978 with the message that their health was seriously threatened. That message, combined with mysterious events (like strange eruptions) drastically changed the life of the residents. Their houses were built upon an old canal that had been used as a deposit for hazardous chemical waste in the forties and fifties.

Sick children sue after playing on waste tip

Fourteen children in a South Wales village are suing the Welsh Development Agency claiming exposure to contaminated land has caused sickness and may have made them vulnerable to life-threatening illness. The children, now aged between seven and 11, live in the same street five yards from a strip of land contaminated with chemical waste from a disused cocking plant in Abercwmboi. Waste from the same source is known to contain substances such as mercury, asbestos, phenols and ammonia, many of which are cancer causing. (...) 'The children used to play on the field and one day they started suffering from terrible mouth ulcers. They covered two teeth. One boy had so many he couldn't swallow his own saliva. They also suffered heavy successive nose-bleeds.' Doctors were baffled. One boy's underarm blisters ate into his armpit and failed to respond to antibiotics.

'We didn't link the site and the illnesses and so we never stopped them playing on the land. There was no fence and the area is very attractive to young children'. says Ms Cook. But following publicity surrounding more than 300 cancer cases among former workers at the nearby coking plant, residents became suspicious. The Welsh Development Agency bought the site behind the houses a decade ago, but has only recently put any fencing around it.

The Independent, February 23, 1997.

Confronted with the health risks and all related social impact, the residents organized themselves to represent their interests. In the following process of decision making they had to struggle for several years with many opponents, not least the authorities and bodies involved in risk assessments.
Exposure to environmental impact is not always as far reaching as in Love Canal and comparable cases of soil pollution. Long-lasting exposure to environmental hazards often results in a decreased annoyance level. This may be due to the fact that some people have moved, leaving their risky or annoying environment, or it may be the result of dissonance reduction (section 7.5.3).
Familiarity with risks may occur in cases of diffuse hazards which are not acute. If pollution obviously does not cause immediate casualties or health threats, the psycho-social impact remains limited to a level people can cope with. Whenever disasters occur, like the reactor accident in Chernobyl, or pollution becomes manifest, as in areas polluted with soil like Love Canal, a totally new situation is created. Then the reactions of people on the location of such an accident appear to be stronger, as was shown in the cases of the near accident in 1979 with the nuclear reactor at Three Mile Island near Harrisburg (Baum *et al.*, 1981; Rankin *et al.*, 1981). Large numbers of people, probably more than a hundred thousand, left the area. Evacuation of an entire area occurred in the small North Italian town Seveso (1976), when dioxins were released into the environment as a result of a serious accident in a chemical plant. Another catastrophe with a large number of evacuees and also with causualties, is the nuclear reactor accident in Chernobyl, Ukraine, in 1986. Examples of catastrophes with large numbers of casualties are two disasters in 1984, one with a chemical plant in Bhopal, India, and another at a LPG-storage facility in Mexico City.
If hazardous substances are found in a district the residents are confronted with a new, deeply encroaching threat which is radically disturbing their lives. The threat is invisible and diffuse, and therefore creates great insecurity. Stress becomes unbearable because the danger is impossible to neglect or to trivialize. At the same time the residents loose control over their own situation and become highly dependant on others. A situation of conflicts and short-term decisions then arises. According to Janis and Mann (1977) behaviour and decisions are determined to a large extent by the pressure people feel. A situation of severe pressure in very risky circumstances not only determines the moment people start to act, but also what they will do. In their theory on decision making in conflict situations Janis and Mann (1977) distinguish five levels:

- ⋆ *unconflicted adherence*. An assessment offers a perspective of limited risk linked to a continuation of normal behavioural patterns. In this situation one does not feel the existence of a conflict, so there is no behavioural change.
- ⋆ *unconflicted change*. The behavioural costs linked to a change of behaviour are not considered as serious, so the conflict situation evokes adapting behaviour, which is creating a situation in which no conflict is felt.

★ *defensive avoidance*. Risks are considered as very serious, but there is no hope of a solution. In this case there is a tendency to avoid the decisional problem, or to transfer it to others, or to postpone it.
★ *hyper-vigilance*. The situation is urgent and requires immediate decisions, but there is no time for seeking good solutions. The result is a kind of panic behaviour, in which people either have strong doubts and frequently change their minds within a short period of time, or in which people stop thinking about alternatives and become focussed on only one particular alternative.
★ *vigilance*. The situation offers sufficient time to come to a decision. In that case a process of trade-off may start and an intensive quest for alternatives can be started and then people can cope with the problem.

How community responses will develop will be determined essentially by the existence of alternatives which are recognized by the residents themselves (also section 8.2.6). If there are no behavioural alternatives, for example the residents cannot move out because their capital is invested in their homes which have no longer any market value, a feeling of strong, helpless anger emerges. The situation of full dependency on others, such as in Love Canal, is an illustration. As soon as the soil pollution is discovered, the residents start to ask the authorities to offer a solution, as they feel themselves captives of the situation (Levine, 1982). A conflict is the result. Extreme helplessness causes anger and mistrust in the authorities because people feel these are a barrier to reaching a solution. The way authorities are handling the problem and managing the risks, may evoke even more indignation than the pollution itself.

The residents also mistrust the information given to them by experts and authorities. This is a result of the divergence between scientific and social rationality. Mistrust of authorities and experts is a common phenomenon in all environmental risk conflicts, in cases of manifest environmental damage as well as in facility siting disputes. It becomes manifest in the fundamental differences between official risk assessments and lay risk perceptions. People confronted with severe manifest hazards feel their questions are not answered by risk technicians and they get answers to questions which were never asked (see section 7.6.2). The answers they get are official risk assessments based on risk analysis. How are risk technicians actually handling risks and why isn't that convincing enough?

7.7 Risk analysis

Authorities, industries and other organizations involved in risk-generating activities try to control environmental risk and annoyance. Risk analysis is the generic term for the efforts directed at the assessment of risk in advance. Anyone who is willing to consider the possibilities of environmental damage resulting from his or her activities, should find a method to balance these negative consequences with the expected benefits. Cost/benefit analysis may be a good method when all costs and benefits can be assessed in an economic sense. When the cost are not economically determined, weighting becomes a complicated process. As a result of the awareness of the environmental impact, the need to incorporate possible future costs in decisions has grown. The basis for these assessments can be found in decisional theory (Fischhoff *et al.* 1987, p.1002). Several attempts are made to design methods and formal procedures of the assessment of potential hazards. Risk analysis methods and procedures are developed to underpin decisions that may lead to the creation of hazardous situations.

7.7.1 Revealed preferences: 'acceptable risk'

Among decision makers there is a strong preference for objectifying methods of risk assessment. Risk analysis is used to describe and possibly to quantify the probabilities and consequences and compare the outcomes. To make it objective the means of description and quantification have to be formalized as much as possible. In these procedures there is a strong desire to formalize the decision that is made with the outcome of the calculations, which implies a decision model in which calculated risks are compared with standards. So the establishment of standards of acceptability is needed for these procedures.
One of the first attempts to develop formal decision making including quantification of risks was made by Starr (1969), whose analysis is basically a utility model in which costs and benefits are weighed. The basic assumption is that for each activity there is a level of risk that will be accepted by the society. That level depends on the utility that is generated by the activity. In Starr's analysis the level of acceptance is not indicated by preferences, but is a figure based on actual behaviour. The fact that people conduct certain activities apparently means that they accept the risks involved. The lack of public controversy about and opposition to activities such as skiing, cycling and conventional electricity generation, obviously means that these activities are accepted by society given the present number of casualties. This *revealed preference* method is the foundation of the risk analysis. Costs and benefits are objective figures: the indicator of risk is the number of casualties per hour of exposure per capita and utility is expressed in monetary units.
From these revealed levels of acceptance Starr tried to extrapolate levels of acceptance for new technologies, in particular nuclear energy. The conclusion of that analysis is that basically there are two curves representing the equilibria of costs and benefits, one for voluntary and and one for non-voluntary risks (Starr, 1969, p.1234). From an analysis of mining accidents a third power relation between risk and utility is derived for describing the curve of acceptance levels. The acceptance level of voluntary risks such as skiing, smoking etc. is thousand times as high as the acceptance of involuntary risks (including environmental

risks), such as electricity generation.

7.7.2 The scientific rationality of risk analysis

As previously discussed, risks appear to be social and cultural constructs. Objectivity is impossible as the valuation of risks is time-, place- and situation-dependant. As noted in section 7.6.1 the fact that society has decided to use a technology, or to build a facility, does not mean that the risk associated with it is acceptable in an absolute sense. As described in section 7.6.1, in all decisions only options can be accepted, not risks, which are merely a feature of options. The basic assumption in risk analysis is that risks can always be weighed against benefits and therefore have a market value. Sagoff states that the value of environmental goods cannot be determined on a market, so neither can environmental risks. Environmental goods are collective goods, so the value of these goods can only be the result of collective decision making. Environmental risks will always be weighted in a political process, not in a trade-off procedure of a technical nature. Environmental risks are not technical problems that can be solved by merely calculating them (Sagoff, 1984).

Standards will still be needed to decide whether or not an option can be accepted. Many risks in society are managed by setting standards, such as standards for emission levels or exposure levels. When using this way of deciding upon activities, the decision process is only moved to the process in which the standards are established. Setting the standards will still be collective decision making and therefore the results of polical processes.

An example of practical impossibility of knowing all risks.

(1) There are about 7 000 000 known chemicals.
(2) Approximately 80 000 are in commercial circulation
(3) Approximately 1000 new chemicals enter commercial use each year.
(4) Using the total of world laboratory resources, about 500 chemicals per year could be testable for toxicity (at colossal expense)
(5) One test, for carcinogenicity alone, can involve 800 test animals and 40 different tissue specimens per animal. This needs approximately $500 000 and 3.5 years
(6) There are approximately 14 000 food additives and contaminants. Many natural components are also thought to be toxic.
(7) The US EPA hazardous substances lists contains approximately 500 substances. In practice only 114 are regulated.
(8) The US EPA's EP test for hazardous wastes covers only 14 chemicals (...)

Wynne, 1987, p.48-49. Excerpted with kind permission of Springer-Verlag © 1987.

Even then, the methods and calculations used for establishing the risks also contain standards for scientific procedures. A basic principle is that an expression of all scientific uncertainty should be part of the outcome of all research. Uncertainty will be great in many cases, and it may even be in principal and in practice impossible to reduce it to a reasonable level, as illustrated by an example (p.164). The example is taken from an analysis of the management of the risks of hazardous waste. In that case a reasonable assessment of risks and scientific derivation of standards are impossible. They cannot be made possible either by intensifying research, so the conclusion is that for chemically hazardous waste the risk analytical approach offers no possibility of solving the problem of acceptability (Wynne, 1987, p.78).

The use of standards based on risk analysis often becomes a barrier in the process of decision making on measures against the source of environmental risks. The effect of defining standards and improving the scientific procedures for risk assessment leads to a paradox. When the standards of scientific accuracy are turned up, the circle of risk justifying actions shrinks. When the scientific standard for accuracy is improved, the probability that a risk is denied grows. A significant problem for example is that health risks mostly cannot be unequivocally attributed to one single source. The cause-effect relation in a case of toxic soil pollution and cancer clusters cannot be scientifically established. The issue here is, whether or not the scientifically established cause-effect relation should be required before legitimate action is taken.

As in this practice risks only exist when they are positively established, the 'necessary counteractions are neglected... To put it bluntly: insisting on the purity of scientific analysis leads to pollution.' (Beck, 1992, p.62). By increasing demands of scientific accuracy the number of cases grows in which the conclusion of a risk assessment should be that the state of information is 'unclear'. Necessary countermeasures will be neglected, as these are not legitimized by scientific results.

7.7.3 Do risk assessments make sense?

Decision making which takes into account environmental considerations, is often called 'environmental management', or 'risk management'. In risk management and decision making about environmentally sensitive projects, the role of the public is of particular interest. As the outcomes of risk assessment will always be biased by the approach being used, the result may be helpful in the process of decision making, but the ultimate decision will always have a political dimension. As a matter of fact many decisions in politics are about the distribution of risks in society, with environmental risks as the most significant category. This is the central thesis of *Risk Society*, a book by Ulrich Beck wich has had a great impact on the discussion about environment, risks, and society (Beck, 1992). He states that contemporary modern industrialised societies should rather be called 'risk societies'. In risk societies decision making and politics is more about the production and distribution of hazards that it ever was. In risk society 'bads' are distributed, whereas the old industrialised society was primarily structured to distribute 'goods' (Wynne, preface in Beck, 1992, p.3).

The way modern societies try to deal with these new problems of distribution will be described in chapter 9. An example of an important instrument that will be discussed there is Environmental Impact Assessment (EIA). In EIA procedures principles of risk assessment and decision making are combined for the management of risks in concrete decisions like siting facilities. As technical risk assessments are still part of this way of risk management, there is still a need to judge the value of these calculated risks. Are calculated probabilities for disasters or health problems caused by environmental damage, good predictions of reality? Much of the environmental risk literature is based on studies in the field of nuclear power. In fact, technological assessment and risk analysis have benefited greatly from the extensive discussions about the use of nuclear technology. For example the revealed preferences approach of Starr, has emerged as a result of the public doubts about the use of nuclear power. In addition to the refinement of methods for risk analysis, a research tradition in risk perception developed, mainly because it was felt that the factors determining the public misconceptions about nuclear energy should be discovered and then corrected.

In other words, risk technicians calculate actual risks, the public is only perceiving risks. Wynne states that scientifically founded risk assessment is only a particular sort of risk perception. That conclusion is drawn based on four case studies, the herbicide 2,4,5-T (containing traces of dioxin), animal growth hormones, hazardous waste disposal, and the risk of an underground methane explosion. In all cases calculated risks did not relate to actual risks established afterwards (Wynne, 1989, p.36).

How can risk assessments be so wrong in the case of many technologies? In his book *Normal Accidents* Perrow shows that risk analysis used for improvement of safety can be effective and useful (Perrow, 1984). However, the main point in his analysis is that there are particular types of accident that cannot be avoided. Probabilities of one in a million per year, which often appear in risk studies, do not exist, according to Perrow. In these risk calculations a particular sort of risk is not considered. The complexity of some systems only allows reasonable assessments of risks of system components. Probabilities of the combined effects of failures can scarcely be calculated, and the consequences of combinations of failure cannot be known at all.

Part of the problem is that complex technological systems do not only contain material components. People are always part of these complex systems, and they may cause completely unforeseen complications. Unforeseen acts of people in combination with technical components may cause surprising effects.

The role of people in risk management appears to be an essential part of public risk perceptions. When people are judging technologies, or planned facilities, or the management of acute environmental damage, they include the human factor in their perceptions. They not only have ideas of the dangers of a new technique, they also judge the reliability of the people and the organizations that have to manage the risks. In a study in the UK on the siting and risks of waste incinerators the dangers people attributed to the facilities were not primarily linked to the technique of incineration. The mistrust was concentrated even more on the industry that was supposed to manage the new facilities. The industry that is involved in waste disposal has an extremely poor image, and its professionalism has also been questioned (Petts, 1992, p.175).

So in the discussion on risks, the public will include their perceptions on the ability and reliability of the organizations involved in creating, assessing and managing the risks. These organizations include normally the companies involved in managing facilities, but the government, public bodies and local authorities as well. In many cases of facility siting these bodies are cooperating with the initiators, and at the same time they performs tasks in controling the facilities. The public also evaluates these inspection and control functions and these are not always performed in a convincing way. Evaluating the people and organizations involved in risky activities is part of the social rationality that is the foundation of public risk perceptions.

7.8 Cultural backgrounds of risk

The differences in the perception of risk have a social background. Personal perceptions are partly dependant of the individual's position within society. The perception of some particular threat is coloured by interests, not only the interests of not being threatened, but also by the interest in the benefits of the activity that is causing the threat. Because risks are always judged within the context of some framework of decision, the judgment of risk will always be affected by the position of the judge in the decisional situation. However, this does not imply that all risk perception is completely individually determined. There are some structural trends in the perceptions of risk and in the literature there is a growing consensus about these general patterns. This consensus contains the notion that cultural perspectives are behind the trends in the judgment and management of risk.

The cultural approach to risks began with the studies of cultural anthropologist Mary Douglas and colleagues. In their analysis risks are not simply a product of impact and probability. They are 'a joint product of knowledge about the future and consent about most desired prospects. (...) The risk debate itself is a record of rapid cultural change. (...) Instead of the old recurrent imagery of knowledge as a solid thing, bounded or mapped out, we prefer the idea of knowledge as a changing product of social activity.' (Douglas and Wildavsky, 1982, p.5 & p.192). People are selective in their awareness of danger and the way they handle these dangers. Within their mind and in their social behaviour people confirm it by their social life. Risk perceptions and decisions concerning risks are therefore fit in their value systems as well as their way of life. The result is that we judge risks from a viewpoint that is related to our position and role within society: 'different forms of social organization are disposed to take (and avoid) different kinds of risk.' (Douglas and Wildavski, 1982, p.9). Hence, risks are *collective constructions*, which becomes manifest particularly in institutions and social organization. In cultural theory strong links between social relations and cultural bias result in a few viable cultural perspectives that are claimed to be characteristic for valuation and handling risks, including the development and implementation of environmental policy (Thompson *et al.*, 1990). Although Douglas' assertion that forms of social organization cause particular ways of thinking about nature and risk, is not accepted by the entire cultural anthropologist community, 'the view that human society provides models for understanding the world is quite widely held.' (Milton, 1996, p.216).

In section 4.8 the approach to environmental risks has already been placed within the frame of four of these cultural perspectives. There cultural theory was used to distinguish between different 'myths of nature', which are linked to archetypes of views on social reality and distinctive forms of rationalities (Schwarz and Thompson, 1990, p.7). The basic social views and rationalities are directly related to the way risks are viewed and handled in decision making. Thompson *et al.* (1990) distinguish between five 'ways of life', each representing a viable combination of social relations and cultural bias, in which the latter is an expression of shared values and beliefs. One of these five cultural perspectives, the *autonomous* perspective of the hermit, in which everything that happens in the world is irrelevant, does not offer a perspective for any kind of action. The other four are (Thompson *et al.*, 1990, p.6-7).

- ★ the *fatalist*, with a fatalist rationality which denies all predictability and control so risks cannot be managed;
- ★ the *individualist*, whose rationality is personal utility-seeking with an attitude of laissez faire, tending to take risks and trying to manage them effectively in order to increase the possibilities;
- ★ the *egalitarian*, who has a more critical view on the needs and is rather looking for ways of preventing than managing risks after taking them;
- ★ the *hierarchist* has a strong procedural rationality towards society in general and risks in particular, so risks should be accepted and controlled by regulation.

Of course, in practice most people do not completely fit into only one of the descriptions. The types represent extreme positions and most people will be located somewhere in between. However, without claiming any general validity, the types appear very illustrative and we can attribute the perspectives to the different positions we have described in the previous sections on community response and to risk analysis and management in the cases of facility siting and obvious environmental impact.

As fatalism and the autonomous perspective will hardly cause any participation in the decision making process, the other three stand for the social conflicts that often arise in these cases. Project planners and technicians generally represent the individualist perspective, as they are optimistic about the management and control of risks of their projects and their are mainly focussed on the benefits of their activities. They will find other individualists on their way, as they meet the *type A* opposition (section 7.6.2). These typical nimby's will only cooperate when a substantial part of the benefits will be theirs. Hierarchists can mostly be found within public bodies that are concerned with control and regulation, in which they often are confronted with the individualist ('if we are limited in our activities, we will go abroad and generate our benefits elsewhere'). The fundamental egalitarian position is in the extreme represented by the typical grassroots opposition, with a preference for avoiding the risk instead of regulating it, because the belief in the knowledge and skills needed for regulation and control are absent. As these perspectives are fundamentally different, communication between proponents of them is often impossible.

Chapter 8
Organizations

8.1 Organizations and environmental improvement

Human activities can be divided into individual and collective acts, and into spontaneous and controlled activities. The concept of 'organization' refers to a collective of cooperating people of which activities are not completely spontaneous but more or less planned and controlled. Organizations can cease to exist. People can start and finish membership of an organization. But it is also possible that an organization survives all its members of a certain moment. Organizations can have widely different goals.

From the viewpoint of environmental protection two extremes can be distinguished. The first part of this chapter will be devoted to environmental organizations. These are organizations which consider environmental protection as their main task. In theory this type could be contrasted with organizations which destroy the environment deliberately. However, hardly any organization will be prepared to declare the destruction of nature and environment as its goal. Sometimes the activities of military organizations can be qualified as environmentally destructive on purpose. (see section 6.4).

Far more common is the destruction of the environment as an unintended side effect of organizational activities. An endless series of examples of this type of environmental effects of organizations could be mentioned, such as air pollution and noise as results of activities of travel agencies and pollution of surface waters and dumping of waste as a consequence of the activities of industrial enterprises. From the wide variety of organizations with environmental side effects we choose industrial enterprises to discuss more extensively in the second part of this chapter. However, our main question will not be why and how these organizations damage nature and the environment but, on the contrary, how they try to limit the disturbing effects of their productive activities.

A third category of organizations that will be discussed below are so-called utilities. Under that term we refer to those organizations which fulfil essential functions like the delivery of energy and the removal of waste. As in the case of industrial enterprises our main question will be whether and how these organizations try to limit environmentally damaging effects of the delivery of their services.

8.2 Environmental organizations

8.2.1 Environmental action

Earlier, in chapter 6 on environmental attitudes and behaviour, we discussed the many factors which often prevent environmental attitudes gaining any influence on environmentally relevant behaviour. In that chapter we were mainly interested in the environmental impact of all behaviour and in the ways in which people can limit the impact of their *own* activities. Examples of behaviour not primarily motivated by environmental concern (type IIIa and IV, section 6.3.2) are household energy saving and the separation of components of household waste. In some cases this type of behaviour is only motivated by environmental concern, such as the use of bicycle, tram and train instead of the private car and airplane as means of transport (type II).

Many political decisions with far-reaching environmental impact are hardly or not at all affected by individual environmentally motivated behaviour. This is the case, to mention just a few examples, with decisions about the construction of nuclear power plants, dams and roads and other infrastructure works, and decisions about permissible levels of environmental pollution caused by private enterprises. In such cases it is not that the behaviour of individuals should be changed, but that the decision-making process should be influenced in order to reduce undesirable environmental effects. The objectives and actions of private and/or public organizations have to be changed. In a broad sense the term *environmental action* is used to refer to all human activities which intend to change behaviour of other people in order to protect the environment.

So environmentally conscious individuals can either intend to change their own behaviour or try to change the behaviour of others (figure 8.1). In the latter case

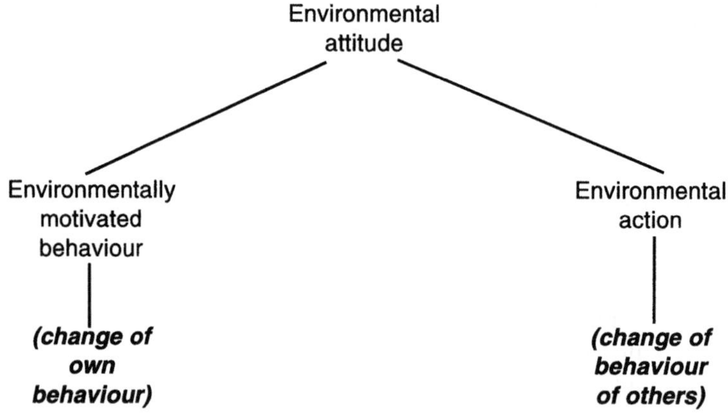

Figure 8.1 Environmental attitude, behaviour and action

this is considered environmental action. Environmental action can be defined as any activity that intends to change the behaviour of others in order to protect the environment.

Between the extremes of environmentally motivated behaviour without any social effect, like the individual consumption of a vegetarian meal at home, and environmental action against political decisions which are not affected by individual behaviour, there are mixed cases in which environmental behaviour has a demonstrative effect as well as the intention to change the behaviour of others, including decision-makers. This is the case with demonstrative environmental action when 'ecological' ways of food production and consumption, energy use and transport are put into practice in order to influence other citizens and political decision-makers to follow these examples and to stimulate the spread of ecologically adapted life styles.

8.2.2 Movement and organization

The concept of *environmental action* refers to human activities on a small and a large scale. It can be used to indicate the activities of a single person who tries to prevent the felling of a tree in front of his house during a short period of time. However, it also refers to the struggle against nuclear energy by many people and in many countries over many years. To indicate the latter type of action the term *environmental movement* is used. Characteristic of any social movement is the massive mobilisation of people in order to reach a common goal. There are peace movements, emancipation movements and there are environmental movements. The concept of 'movement' refers to the fact that people perform activities which do not belong to the more or less fixed behavioural patterns of their normal social roles as family member, student, worker or member of any organization. Demonstrations, blockades, signing petitions and wearing special clothes belong to the action repertory of social movements.

The term (environmental) movement is often used as a synonym for one or more (environmental) organizations. However, it makes sense to distinguish between these two types of collectives. There are large tourist organizations, but the term 'tourist movement' is seldom or never used to refer to these organizations. The reason is that the members of tourist organizations pay for their membership but are not 'moved' to further activities to defend the interests of tourists. In the same way people can limit their membership of environmental organizations to the payment of their annual contribution. In that case they are members of an environmental organization but do not belong to an environmental movement. The use of the term environmental movement should be limited to those situations in which the members, and not only the professional staff, of a group are actively working to reach common group goals. This distinction between 'movement' and 'organization' does not exclude the coming together of environmental organizations and environmental movements. On the contrary, environmental organizations can initiate massive mobilisation of people in order to reach common targets. On the other hand, more or less spontaneous movements will often only be successful if their activities are coordinated by an organization. As we shall argue below, in some cases massive mobilization of environmental activists will support the work of an environmental organization, in other cases it will be irrelevant and in still other cases it may even damage the goal attainment of environmental organizations.

Notwithstanding the presence of large environmental organizations operating world-wide like Greenpeace, small and locally limited organizations are still continuing their existence and even new ones are being founded. Our definition of environmental problems (see chapter 1) implies that new human activities and new opinions about the acceptability of environmental effects of human activities continually create new environmental problems. Individuals can always take the initiative to found an environmental organization to counteract human influences on the environment which they consider as being problematic. In doing so they can be inspired by widely different values and goals like the striving for individual independence and the preservation of the national culture (Yanitsky, 1993, pp.30-39).

8.2.3 Strategies and methods

Environmental organizations try to reach their goals by means of different strategies and methods.

The strategies of environmental protection organizations can be divided into two broad categories: influencing individual and institutional behaviour (section 8.3.1.) and influencing political decision-making (section 8.3.2.).

8.2.4 Influencing individual and institutional behaviour

Part of the activities of environmental organizations are directed towards individual citizens, private enterprises and other non-governmental organizations. An important part of the activities of environmental organizations consists of the spread among citizens of written, visual and oral information and argument. Many environmental organizations have their own magazines and often have easy access to mass media such as press, radio and television. Other activities in these area are exhibitions, excursions and demonstration projects on ecological farming or housing.

Organizations with environmentally damaging activities are approached in different ways: in an indirect way by influencing governmental decision-making and in a direct way by making agreements with private enterprises about the reduction of their environmental pollution without the involvement of governmental authorities.

8.2.5 *Influencing political decision-making*

Many forms of environmental damage can only be stopped or prevented by the decisions of organizations, and in particular by governmental authorities. For that reason environmental organizations invest much energy in influencing the decision-making processes of governmental authorities, from a local to a world-wide scale. This strategy can be put into practice by means of different methods which can be divided into argument and mobilisation and in the use of mental and physical means of action.

– Argument and mobilisation

Forms of *argument* frequently used by environmental organizations are lodging objections, taking matters to court or using other opportunities offered by the judicial system and lobbying political authorities, members of parliaments and other political councils as well as civil servants.

Examples of *mobilization* are the collection of signatures to support a protest action and the organization of demonstrations or blockades. The choice between argument and mobilisation is the choice between the quality of the argument and the quantity of the support. Both, of course, can be combined. Examples of the combination of argumentation and mobilisation are forms of public information like exhibitions and excursions.

– Mental and physical means

Often, non-violent and violent means of action are distinguished. However, the difference between these two categories is not always clear. The blockade of the entrance to a nuclear power plant may be called a violent means of action by some people and non-violent by others. Therefore we prefer the distinction between *mental* means of action of which the primary aim is to convince people and *physical* means of action of which the primary aim is to disturb other people's activities.

To the mental means of action belong the above mentioned different types of argument and forms of mobilisation like collecting signatures for a petition and mass demonstrations.

Physical means can be used by mobilized masses but also by individuals or small groups. The spectacular actions of Greenpeace to stop whaling and nuclear tests are examples of the use of physical means by a small number of well-trained professional activists. In the past, blockades of nuclear power plants in Germany were examples of the use of physical means by mobilized activists.

Type and size of environmental action cannot be explained by the degree of environmental disturbance caused by human behaviour. In some cases drastic forms of human induced environmental change will not provoke any environmental (re)action, whereas in other cases comparatively minor human effects on the environment will meet with strong environmental protests. Research of anti-nuclear protests confirmed the 'absence of a direct relationship between objective conditions, grievances, and mobilisation' (Kriesi *et al.*, 1995, p.162).

Argument and mobilisation can go together as action methods within one organization and even within one actual action. Physical and mental forms of action are more difficult to combine. Greenpeace is an environmental organization which uses both mental means like lobbying and judicial procedures and physical means to prevent whaling and dumping of nuclear waste. But time and again there are tensions within Greenpeace between these different forms of action. The use of physical means of action can damage vested interests of professional environmental organizations. It can have negative effects on fund raising, recruiting of membership and on lobbying among politicians. Taking personal risks in violent actions requires a different organizational culture from lobbying and going to court.

8.2.6 Coalitions and conflicts

The public support of environmental actions depends on the personal interests of citizens. Often there are conflicts of interests between environmental protection on the one hand and all sorts of personal interests of citizens on the other hand. A notorious example of such a conflict of interests is the conflict between 'employment' and 'environment'. In theory environmental protection can be combined with full employment. The introduction of a so-called 'ecotax' has been defended as a first step in making the use of raw materials and energy more expensive and the use of manpower cheaper. In this way it would contribute to both the creation of jobs and the reduction of the use of scarce natural resources. However, when the operation of one single factory is at stake, there are often conflicts of interests between employees and employers who want to continue or even to extend their productive activities on the one hand and environmental organizations who want to limit or even completely ban these polluting activities on the other hand.

Some forms of environmental disturbance are extremely damaging to the personal interests of citizens. Surface water pollution may damage the interests of fishermen. In developing countries in particular many examples can be found of disturbances of ecosystems, like tropical forests, on which native peoples are dependent for their complete livelihood. For example the construction of dams in order to produce hydropower may lead to the disappearance of areas on which local inhabitants were completely dependent.

Besides situations in which there is either a positive or a negative relationship between environmental and personal interests, there are situations in which personal interests are neither positively nor negatively involved in environmental issues. This is, for example, the case when a world-wide organization like Greenpeace fights environmentally disturbing activities in other parts of the world. In countries which have no vested interests in whale hunting people will neither take profit nor suffer damage from actions against whale hunting.

Martyrs of environmental protection

Chico Mendes

On the night of 22 December 1988 Chico Mendes the leader of a Brazilian trade union of rubber tappers was assassinated. Mendes had been highly successful in initiating actions to stop the clearing and burning of forests in the Amazon. Mendes' biggest triumphs came both home and abroad where he was championing his cause under the umbrella of 'environmentalism'. He had enjoyed a rapid rise to both national and international fame and Chico Mendes had become the human face of the fight to save the rain forest. With Mendes' involvement with the rubber tappers, the authorities could no longer dismiss efforts to save the rain forest as foreigners interfering in Brazilian affairs (Rowell, 1996, p.214).

Ken Saro-Wiwa

On 10 November 1995, Ken Saro-Wiwa, the leader of the Movement for the Survival of Ogoni People (MOSOP), and eight other Ogoni were hanged by an illegal military junta in Nigeria after a trial condemned by the British government as 'judicial murder'. Saro-Wiwa's real crime was to highlight to the world the ecological devastation caused by the multinational oil company Shell and to campaign for a greater share of the oil wealth that had been drilled from under Ogoniland. The Ogoni paid the ultimate price for fighting for environmental and social justice. Over 1,800 have been killed, 30,000 made homeless, and countless others imprisoned, raped, tortured and beaten up. (Rowell, 1996, p.1)

Another factor which influences the reaction of people towards environmental actions is the difference between actions against already existing and actions against expected future situations. People may become adapted to all sorts of environmental disturbance. If they do not perceive a real chance of amelioration of their situation they will tend to protect themselves against the illusion that improvement is possible. For that reason they may reject environmental actions and regard prevailing environmental circumstances as satisfactory. This tendency to protection against disillusionment explains the fact that sometimes the people which are the most exposed to some form of environmental pollution are the least inclined to protest and deny or play down the annoyance caused (section 7.5.3). This type of minimisation or denial of suffering from environmental disturbance is less to be expected in the case of environmental protest against future disturbances as a consequence of projects which are only planned or under preparation. In such cases more support may be expected from people who have no personal interests in the proposed project.

The comparative study of anti-nuclear protest in different countries taught that there is far more opposition against planned nuclear power plants than against plants which are already in operation. 'There has been very little opposition against existing nuclear installations, the vast majority of attention of protest is directed towards *planned* nuclear capacity' (Rüdig, 1990, p.107; also see section 9.3.1). Figure 8.2 combines the dimension of common versus conflicting interests and the dimension of actual versus future disturbance.

	Existing disturbances	Future disturbances
Action supports personal interests	Fishermen in case of actions against pollution of surface waters	Local citizens in case of protests against airport extension
Action does not interfere with personal interests	Citizens of countries without seals in case of protests against seal-hunting	People of developed countries in case of protests against the construction of dams in developing countries
Action hurts personal interests	Employers and employees of enterprises which are targets of environmental actions	Road constructors in case of protests against the construction of new roads

Figure 8.2 Examples of different personal interests that are involved in environmental actions against existing and future environmental disturbances

8.2.7 Nature conservation and environmental protection

So far we have neglected the distinction often made between environmental protection and nature conservation. What was said about environmental organizations in general also applies to nature conservation organizations. Many organizations (including Greenpeace) can be considered as environmental and nature protection organizations, yet it makes sense to distinguish specific characteristics of nature conservation organizations from those of environmental protection organizations.

Typical nature conservation organizations in general have a longer history than organizations which are primarily active in the field of environmental protection. As early as the end of the nineteenth century and the beginning of the twentieth century nature conservation organizations were founded in western Europe and northern America. In France in 1854 the 'Société impériale zoologique d'acclimatation' was founded in order to protect the existing fauna and to enrich it by means of the introduction of exotic animals. In England the 'Commons, open space and footpath preservation society' was founded in 1867 and in the USA the National Audubon Society which still exists was founded in 1905.

Table 8.1 The big ten environmental organizations in the USA.

Organization	foundation
The Sierra Club	1892
The National Audubon Society	1905
The National Parks and Conservation Association	1919
The Izaak Walton League	1922
The Wilderness Society	1935
The National Wildlife Federation	1936
The Defenders of Wildlife	1947
The Environmental Defense Fund	1967
The Friends of the Earth	1969
The National Resources Defense Council	1970

These and other nature conservation organizations have in common that they want to protect pieces of nature which are of special value because of their beauty, rareness or scientific interest.
Environmental protection organizations in general have a shorter history. A well-known organization with branches in many countries is Friends of the Earth which was founded in San Francisco (USA) in 1969 (See also section 10.7). Unlike nature protection organizations, environmental organizations do not strive for the protection of exceptionally valuable parts of nature but for the maintenance or restoration of environmental quality in general. Highly simplified the difference can also be summarized as follows: the main targets of nature conservationists are valuable pieces of nature and the main target of environmental protectionists are polluting organizations and individuals. Related to this difference is a difference in style between these two types of organizations. Nature conservationists have a positive message: look how beautiful, how special and how interesting the pieces of nature are which we care for. Environmental activists have a negative message: look how dirty air and water surrounding that factory are and how terrible these waste dumps are (Hertz and Tellegen, 1995).In addition to the strategies and methods discussed above nature conservation organizations try to achieve their goals by the acquisition and management of nature reserves. The exclusive right to visit these reserves sometimes is the main reason for becoming a member of a nature protection organization.

8.3 Industrial enterprises and environmental improvement of production

8.3.1 Production processes and products

Ask anybody about the main source of environmental pollution and 'industrial production' will probably be mentioned first. From the start the development of industrial production has been accompanied by complaints about pollution, noise and risks. In the recent past negative public attitudes to industrial enterprises were strongly influenced by environmental catastrophes caused by industries like the methyl isocyanate gas leak in Bhopal (India) in 1984, which caused 3000 deaths and 250.000 injuries within the first week after the accident (Bennett, 1992, p.194) and the release of chemical pollutants in the river Rhine as a consequence of an accident at the chemical plant of Sandoz in Switzerland in 1986.

However, it cannot be denied that industrial enterprises do try to reduce the negative environmental effects of their production. In chapter 3 on geography we discussed growing efficiency in production processes in contemporary market economies. The concepts of 'ecological modernization' and 'de-materialization' were introduced and explained in that context. Environmental improvements of industrial production can be distinguished in improvements of production processes and products. In general, environmental improvement of production processes precedes environmental improvement of products.

Environmental improvement of production processes and products

Organizing for the systematic environmental improvement of products is as a rule more complex and more difficult than organizing for systematic improvement of production. Complexity and difficulty are closely connected with the relatively large number of actors who would normally be involved in successful product improvement. Successful 'greening of a product' often implies coordination between companies involved in different stages of the product life cycle, and close co-operation between different departments within one company, such as for instance the sales/ marketing department, the purchasers of supplies/inputs and the design department. Also the information involved is often more complex than in case of the improvement of a production process within one company.

Reijnders, 1996[b], p.28.

In following sections we will not always strictly distinguish between the improvement of production processes and products. Our main question is why and how enterprises try to reduce environmental stress caused by their activities.

8.3.2 External pressure

Industrial enterprises are highly dependent on the support of their social environment. All sorts of organizations and groups can put pressure on enterprises to reduce their emissions and other forms of environmental stress. Williams *et al.*(1993, p.123) mention the following interest groups as possible sources of external environmental pressure on enterprises:
government; employees; suppliers; customers; investors and the local community.

As part of their environmental policy *governmental authorities* at different policy levels can put pressure on industrial enterprises. In chapter 9 on state and policy we will discuss different instruments of environmental policy which can be used by governmental authorities in order to force industrial enterprises to comply with their environmental standards.

Employees may require environmental measures of their employers because such measures will have beneficial effects on their health. Another reason why employees may be interested in such measures is the longing for 'their' factory to have a better reputation. In general employees dislike being blamed by family members and friends for working in a polluting or dangerous factory.

Suppliers of consumer goods may prefer to sell products of factories with a good environmental reputation and *customers* may prefer to buy such products. *Investors* may opt for 'green investments' because they are convinced of the market perspectives of products from enterprises with a good environmental reputation. Another reason for doing so - in particular in the USA- may be liability for environmental damage (Williams *et al.*, p.122).

The *local community* has a direct interest in environmental improvements of nearby industrial enterprises. *Environmental organizations* should be added to the categories of interest groups mentioned by Williams. These organizations can put considerable pressure on enterprises to adopt environmental measures. Whether one or more of the groups mentioned above will put pressure on enterprises to ameliorate the environmental quality of their production processes and whether this pressure will be successful depends on many social circumstances, among which the employment situation is extremely important.

8.3.3 Internalization of environmental control

Both from a technological and a sociological perspective the development of environmental measures by industrial enterprises in rich western countries can be characterised as a process of internalization.

The 'internalization' of environmental technology can be illustrated by the development of technological measures to reduce air pollution. In the past high stacks were built to dilute and to spread industrial air pollution. The next step was the introduction of so-called end-of-pipe technology. Filters were installed in stacks and air pollution was transformed into solid waste. The third phase in the development of clean(er) technology was the introduction of clean(er) production processes, combined with the use of clean(er) materials within production processes. Corrective technology at the end of the production process was replaced by preventive technology within the production process.

Preventive measures can be divided in 'reduction at source' and 'internal recycling' measures.

Two ways of prevention of waste and emissions within industries

SOURCE REDUCTION: minimisation of the amount and/or harmfulness of waste and emissions at source by:
- change of product,
- changing the process technology and the process itself, including change in input of raw materials/additives.

INTERNAL RECYCLING: the recycling of waste products and emissions in which these are
- recycled as raw material in either the same or a different production process,
- processed with the intention of recovering material, or
- used for a different useful approach

Dieleman *et al.*, 1991, p.4.

As a social process the development of cleaner production in recent years can be defined as the transfer from a defensive or compliance approach to an offensive or innovative approach. As long as a compliance approach prevails the necessity to reduce emissions in order to obey fixed standards of governmental environmental policy will be the starting point for environmental measures within enterprises. Ways will be looked for to conform, with the least costs and effort, to the standards formulated by others. When an innovative approach is followed the starting point is the design of production processes which best fit both economic and ecological criteria. In the past decades great successes with such an innovative approach been reached in many countries (Groenewegen *et al.*, 1995; van Berkel, 1996). In many cases it was possible to take preventive measures that created economic-ecological win-win situations. 'Pollution prevention pays' was the principle on which many environmental innovations in industries were based.

8.3.4 Internal and external obstacles

Notwithstanding the positive financial results of many waste prevention projects in industries, such projects often meet with all sorts of resistance within enterprises. One of the obstacles is the position of internal environmental protection activities within the organization. 'Environmental protection systems formalized at an earlier date still appear to be staff-oriented. There is often little or no translation into tasks and responsibilities for the line organization (executive management). Moreover, there is always the possibility of responsibilities being shirked: only one person in the company is responsible for environmental issues and consequently no one else needs to focus attention on such matters. This often causes the organization of the company to have an inhibiting effect on the introduction of prevention concepts as they require an integral approach and often demand the input of both staff and line expertise ' (Dieleman *et al.*, 1991, pp.

59/60). In addition to internal organizational obstacles there are external obstacles to clean(er) production as well (Dieleman *et al.*, 1991, p.64). One of them is a lack of cooperation between companies in the same production column. Many products are the result of several production processes in different enterprises. Environmental improvements require the cooperation between different enterprises involved in the production process (Cramer and Schot, 1993). Another external obstacle concerns governmental environmental policy. A compartmentalized structure of environmental policy in which measures regarding air, water and soil pollution are not integrated will stimulate end-of-pipe technology which translates one type of pollution into another one and will hinder preventive measures. On the other hand integrated policy measures including different types of pollution and other forms of environmental stress will stimulate preventive measures.

8.4 Utilities and demand-side management

8.4.1 Old and new goals of utilities

In modern societies supply and demand of goods and services are generally regulated by the market mechanism. However, modern states have taken the responsibility for providing goods and services which are considered to be essential to human well-being. The provision of drinking water and energy to households and the removal of their solid waste and human excrement are societal functions for which modern states are, more or less, responsible (section 9.2.2). In this section we limit our description to the provision of energy to households and the removal of their solid waste. Organizations have been developed to provide these goods and services, which are usually called *utilities*. In this century in most countries they have been brought under state control, in an indirect way by state regulation, in a direct way by nationalisations or the development of public bodies or even state enterprises.

Reliability at any moment, availability everywhere and affordability for all citizens became the main criteria for the task-fulfilment of these utilities. In recent decades energy conservation and waste minimization have been put high on the environmental political agenda. Measures have been taken to stimulate the reduction of both energy consumption and waste production. State-owned and state-regulated bodies which supply energy and remove waste may counteract policy measures for reducing energy consumption and waste disposal by providing and extending their facilities. Under which conditions can these organizations contribute to waste reduction and energy conservation, in particular the minimisation of the use of non-renewable resources?

8.4.2 Supply-side and demand-side management

The old tasks of organizations responsible for energy supply and waste removal required 'supply-side management': the ability to fulfil, at present and in the future, customers needs for energy consumption and waste removal. The new policy aims of energy conservation and waste minimisation require '*demand-side management*' (DSM): the ability to stimulate clients to change energy consumption and waste production, which may include reduction of energy demand and of waste production. Supply-side and demand-side management require widely different abilities from organizations. The main challenge of supply-side management is to develop and manage facilities of the highest technical quality at the lowest possible costs in order to fulfil customers needs under all circumstances. In DSM the challenge is to communicate and co-operate with customers in a way that changes their behaviour. 'In energy supply, technology is the central concern, but in energy efficiency, contract negotiations and financing are critical to success' (Flavin and Durning, 1988, p.51).

Organizations with a supply-side origin will have difficulties in switching to a demand-side policy. They started to fulfil certain goals once and are marked by a tendency to continue their activities and to stimulate needs for the results of their activities, no matter whether they are publicly or privately owned. Starbuck (1971) mentioned different factors contributing to the striving for organizational growth and distinguished the following 'motives for growth': organizational self-realisation; adventure and risk; prestige; power and job security; executive salaries; profit; costs; revenues; monopolistic power; stability; survival.

Electricity utilities for example, either public or private, do not have a natural vested interest in reduction of electricity consumption. It reduces utility revenues which in turn causes higher consumer prices (Johansen and Hoog, 1995). The same holds for the application of renewable sources (private wind turbines or photovoltaic cells) or highly efficient conversion techniques (co-generation, heat-pumps) by the clients of utilities. This raises the question whether or not energy conservation and efficient end-use can be stimulated effectively by the energy supply companies.

In particular, when the new tasks of utilities imply that the demand for their services should be reduced, the existing organizations and the structure of the sectors in which utilities are operating will probably not offer the best conditions. For electricity generation some parts of the energy efficient production may be interesting for existing parts of the sector, because it offers them opportunities for growth (Slingerland, 1997). Co-generation for example requires the building of new power stations and increases the supply of energy with the distribution of heat or hot water. Application of renewables, in particular by utility clients, and demand reduction are less interesting for existing organizations.

In the sector of waste disposal, in some countries strong impediments for source reduction exist. Vested interests in disposal, in particular high investments in waste incinerators, are creating a barrier to waste reduction (De Jong and Wolsink, 1997). Once incinerators are built, they start attracting waste, because otherwise they cannot be exploited cost-effectively. These vested interests are growing even stronger, because in many countries the interdependency of the waste and energy sectors is increasing fast. The rise of waste-to-energy facilities is one important factor. The fast growing involvement of energy utilities in financial investments for waste facilities is another significant trend (Benzler *et al.*, 1995). The market shows a tendency toward incineration instead of reduction and recycling, which includes a strong pressure (lobby) in most countries to 're-define' energy recovery as recycling (Gandy, 1994, p.114).

If the old organizations and utility-sector are not fit to implement demand reduction and efficient end use programmes, what changes would improve it? In which ways can the old task of supply-side management and the new one of DSM be combined? Two possible ways are organizational integration and task-differentiation.

8.4.3 Organizational integration

Particularly electricity utilities have experience with Demand Side Management and some of them actually achieved energy savings (Nadel, 1992). However, most of the DSM activities were directed at efficient use of installed capacity and grid capacity. Utilities try to diminish differences in electricity consumption between periods of high and low demand (load management), which includes reduction of peak demand, but in practise this often results in stimulating sales during low demand periods. Electricity utilities are interested in this type of demand-side management, because electricity cannot be stored economically on a large scale. Production capacity has to be as large as the peak-demand requires. The smaller the difference between peak and average demand, the more efficient production capacity can be used and the lower production costs will be. Most of the DSM programmes are still directed at the time of use, rather than the amount of consumption (Vine, 1995). However, in recent years electricity producers have also started to take measures to reduce total electricity consumption. Why are these producers motivated to sell less instead of more of their own product? Incentives to do so may have different sources: the market, government regulation and public opinion.

– *Market*

Electricity producers may have a financial interest in the postponement or definite avoidance of the construction of new utility plants to fulfil a growing need of electricity. This will be the case when the costs of electricity produced by new facilities exceed the production costs of existing facilities (Nadel, 1992). Hydroelectric power may have reached its limits and additional production may require the use of much more expensive energy sources. Environmental legislation may require expensive cleaning devices in new facilities which will lead to high production costs. Under these circumstances production growth leads to higher production costs which electricity producers may try to avoid by investing in 'NegaWatts' instead of MegaWatts (Von Weiszäcker *et al.*, 1995, p.194). For this reason DSM may even become big business. Since the late 1980s it has been 'the fastest growing segment of the US power business, reaching an estimated US $ 2.8 billion in 1993' (Flavin and Lenssen, 1994, p.1033).

– *Government regulations*

The USA governmental regulation, based on the principle of 'least cost planning', obliged electricity producers to take measures to reduce electricity demand in cases where the costs of measures to reduce demand were lower than the costs of measures to extend supply. 'One of the most ambitious utility plans is the regional Northwest Conservation and Electric Power Plan, created by federal legislation in the early eighties. It requires that in the future energy efficiency be considered along with new power plants' (Flavin and Durning, 1988, p.52).

– *Public opinion*

The production of energy and the disposal of waste both require large facilities which are often considered as locally unwanted land uses (see section 7.6.2). The opposition against new facilities may cause so many problems, that it may create secondary objectives for promoting electricity conservation or waste reduction. 'The first thing to do in a siting process is to convince people that you really need the power plant and that every possible means of conserving electricity has been considered.(...) Another motive is to keep the electricity sector as free as possible from involvement by politicians and authorities of different kinds.(...) As conservation of electricity is of general interest, it could be very tempting to start dealing with details in this area in order to gain votes' (Jacobsson, 1991).

Demand-side management in waste processing has not been extensively studied and discussed until now. However, in this field also examples of public bodies that both run waste incineration plants and stimulate recycling of different components of waste can be found. Local authorities may introduce waste reduction programs to convince the local population of the necessity of accepting the construction of a new waste incineration plant in their locality (Ehrenfeld *et al.*, 1989).

Practice has taught that market, government regulation and public opinion may stimulate organizations with a supply-side management orientation to develop demand-side management activities. However, one may wonder how stable the integration of demand-side activities in mainly supply-side oriented organizations will be. New developments may make investments in new facilities economically attractive, government may become less interested in reduction measures and public opinion may change. There is no built-in guarantee that organizations with a mainly supply-side orientation will persist in demand-side activities under changing external circumstances.

8.4.4 Task-differentiation

Instead of integrating supply- and demand-side management in the same organizations, different organizations may be developed to fulfil these tasks. In another sector of society, a comparable task-differentiation has, at least in some countries, a long tradition. The dilemma of providing facilities and limiting the use of them as described here for energy distribution and waste removal is basically the same, if not more urgent, in health care. Medical aid should be available at a reasonable price at any time and at every place when it is really needed, but medical consumption and, in particular, hospitalisation should be limited, not only because of economic reasons. In particular, psychiatric hospitalisation may have negative effects on patients. Support-systems may be destroyed and the re-integration into society may be hindered by stigmatisation and hospitalisation syndromes of patients in psychiatric hospitals. In some countries one of the answers to the dilemma of how to offer necessary help and to avoid excessive use of help in health care has been the 'echeloning' of facilities. Patients are expected first to visit a general practitioner before receiving specialist medical care whether outside or inside hospitals. General practitioners have no vested interests in continuity and growth of hospital facilities. They are expected to refer patients to specialist medical care when it is needed, and to avoid special treatment when it is not necessary or even damaging to patients. Such an organizational arrangement may also be useful in solving the dilemma of providing energy and waste facilities as far as really needed yet limiting their use as far as possible. In the case of energy this would imply that the tasks of production of heat, light and power, on the one hand, and their distribution to consumers, would be divided between separate organizations. In the case of waste, a comparable task-differentiation could be made between organizations that manage incineration plants and landfill sites and organizations that remove waste.

There is some recent experience with task differentiation between 'production' and 'distribution' of electricity in the Netherlands. The 'distributors' are now taking all sorts of measures to stimulate energy conservation, but it is far too simple to relate all conservation activities exclusively to the organizational differentiation of productive and distributive tasks (Tellegen and Gilijamse, 1994). However, institutional task-differentiation between supply-side and demand-side management is a model that deserves to be tried and it merits further research.

Chapter 9
State and environmental policy

9.1 The nation-state and the environment

In discussing environmental problems, many people advocate an authoritarian state with far-reaching powers. In their view, such authority is necessary in order to deal with social dilemmas, and to break the pattern by which environmental disadvantages are passed on. The famous article 'The tragedy of the commons' by Hardin (1968), often quoted in the environmental literature and mentioned in section 5.4.1, is a case in point. But are we to assume that an authoritarian state will always make use of its far-reaching powers to solve environmental problems? Are such ideas in keeping with the core tasks of the modern state? Do these tasks include the conservation of the environment?

In calling for the powerful authority required to tackle environmental issues, people easily assume that a vigorous, fairly authoritarian state will use its powers to combat environmental pollution. They see the State primarily as an arbiter whose task it is to settle disputes, and who is, moreoever, in a position to impose its decisions upon society. It is the State which forces all the parties to agree on means to conserve the environment and to overcome conflicting interests. In this sense, the State is the embodiment of 'mutual coercion, mutually agreed upon' (Hardin, 1968, p. 1247). But does the State actually take on the role of an umpire passing judgment on individual interests conflicting with other interests, particularly common interests? The key question is whether or not the State is fully able to represent the common good or the 'general interest'.

9.2 The functions of the modern state

Some enlightened theorists like John Stuart Mill already accepted the environment as a high government priority more than a century ago. Today it is almost universally accepted that the State does indeed have a task when it comes to the environment. But at a time when in many countries the State withdraws to play a more limited role on a variety of fronts, or in any case avowing that that is its intention, it is clear that there is no concensus on what the role of the State should be in the conservation of the environment.

We are assuming that decisions on the best way of dealing with environmental problems are part of a political process, and that the State has an important part to play within that process. However, this role cannot be described simply as 'strong' or as 'the umpire' that is breaking prisoners' dilemma's (Ostrom, 1990). The modern nation state has several roles to play at the same time, reflecting different functions.

John Stuart Mill on environment and authority

Is there not the Earth itself, its forests and waters, above and below the surface... These are the inheritance of the human race.. What rights, and under what conditions, a person shall be allowed to exercise over any portion of this common inheritance cannot be left undecided. No function of government is less optional than the regulation of these things, or more completely involved in the idea of civilized society.

John Stuart Mill, Collected works vol.3, 1965, p.801.

In the past, interventions by government have been primarily reactive in almost any field. The nation-state and its responsibilities are the product of a gradual process of reaction on developments within society. As a result of that process, the modern state now exercises a number of functions which originated in reaction to 'crises' within society. In the order of their appearance, the following functions of the State can be distinguished:

- the structuring function
- the achieving function
- the steering function
- the arbitrating function.

Government policy with respect to environmental problems must be seen in the light of these four functions.

9.2.1 The structuring function

Structuring is the oldest function of the State. From the Renaissance to the Enlightenment, scholars theorizing about the role of the State gradually relinquished the religious element, making way for the first 'modern' theories. The ideas on the role of the State put forward by Machiavelli (1469-1527) and Hobbes (1588-1679) were absolutist. Machiavelli maintained that the authority of the State was not restricted by ethical, judicial or religious considerations. Any means was justified when it came to preserving and strengthening the power of the State, because this was in accordance with the structuring function. In Hobbes' view, the State was founded upon a 'voluntary' social contract. Without such an agreement, all would suffer, for we would have an uncivilised society,

in which each individual was free to do what he liked, even at the expense of others. Hobbes saw the State as an absolute instrument to which its citizens were totally subject; he believed that this was necessary in order to ensure a structure that would do justice to everyone (also: section 5.4.2).
A century and a half later, alongside the theory of absolute authority sketched above, there was the view of John Locke (1632-1704), who attributed an instrumental role to the State. In the way the State structured the life of its subjects, a structural obligation of the State towards its subjects was added. It had not only rights, but also duties with respect to the physical welfare of the country and its residents. The State should also be dedicated to creating a structure that guaranteed the rights of its citizens. For that reason its tasks were limited to those of a monitoring nature: upholding the law, organising the country's defences, and protecting property. This concept is often referred to as the 'night watchman's state'. The structuring function encompasses the classic tasks of the night watchman's state. The protection of property has become particularly significant recently regarding issues concerning the environment. The concept of liability has even become an important instrument of environmental policy these days (section 9.6.2).
Locke's theory became the basis of liberalism, not least because in his view the protection of property was the primary task of the State. In the eyes of the liberals this provided the best conditions for maximum industrial development and trade. This in turn led to the formulation of maximum development as the primary objective of all the actions of the State, and ultimately to the achieving function.

9.2.2 The achieving function

Achievements by the state primarily concerned hardware for economic development that might be viewed as collective goods. In other words, the type of 'hardware' such as bridges, that would not be realized by leaving it to the free market. The largest achievements of the state became all kinds of infrastructural facilities such as roads, waterways, bridges and ports. In the next stage, however, the achieving function started to reflect the responsibility of the State for the social security of its citizens. This also includes collective facilities, not only for the security of the citizens but for the achievement of education and housing as well.
The major obstacle to economic development of the industrial societies at the end of the 19th century was not of an environmental nature, but it was rather a matter of social issues. As a result of the industrial revolution, a steadily increasing portion of the Western population was employed. This resulted in a completely new social structure, one which was characterised by tensions relating to class differences. This problem gradually proved to be an obstacle to future development. The abolition of social obstacles to continuing economic development made way for the welfare state. A number of the problems relating

to the social question were of an environmental nature.

The State may be viewed at as a conglomerate of compulsory, collective arrangements designed to remedy and control the external effects of shortages and setbacks. Combining the theory of collective action (Olson, 1971) and the civilization theory of Elias (1978) De Swaan shows how compulsory structures are the result of a process of collectivization triggered by the need to solve certain social dilemmas. He describes the collectivization of care-taking activities by the state as a civilising process (De Swaan, 1987). Collective facilities, consisting of 'hardware' (infrastructural facilities) and 'software' (organizations, utilities) were created for the provision of collective goods such as the supply of energy and water. Some of these collective goods had qualities in terms of environmental hygiene. In the cities pollution and malnutrition led to disease, which interfered with the availability of workers for the production process. It also resulted in epidemics which formed a direct threat, not only to the working class but also to the wealthy. The concentration of poverty and the overcrowding in low class neighbourhoods created threats in terms of lack of hygiene and safety. In particular sewerage, waste collecting and the distribution of drinking water became part of the achieving function, as it was the State, with the abilities to enforce cooperation and to charge levies for the services, which took charge of these collective arrangements, mostly as a result of the activities of well organized pressure groups. In this way the achieving function of the State was extended beyond purely economic areas. Its responsibilities now included a number of tasks related to environmental hygiene.

9.2.3 The steering function

The economic crisis of the thirties necessitated certain drastic interventions on the part of the State. Its task was no longer limited to providing basic social safeguards and creating the necessary conditions for production. As described in the economic theory of John Maynard Keynes, the State had its own very crucial role in the economic process. To overcome the economic depression the State developed a steering function mainly intended for directed stimulation of the economy.

Economic development, by means of incentives and direction, became one of the immediate aims of the State. A variety of instruments, such as regulations, levies, subsidies, guarantees and other process-influencing instructions, were developed to further that aim. The tasks of the State were considered primarily in the light of what the authorities could do to promote a certain optimum level of prosperity.

The primary concern of governments with regard to economic development became the fight against unemployment. In effect all the major political movements have found the theoretical basis for this view in the economic theories of social welfare: what is economically beneficial is considered socially beneficial. The number of jobs is the primary concern of governments in modern industrialized nations. Therefore the achieving and steering functions of the State are primarily directed towards the creation of new jobs and the maintainance of existing employment. The fact that economic considerations occupy a central position in the operation of the modern state has certain consequences for the manner in which environmental issues are addressed, not only as a result of political priorities, but also in the form of many institutional barriers for the implementation of environmental policy measures.

Many institutions in modern industrialized societies are partly based on state legislation and therefore partly determined by policy choices in the past. Institutions are the working rules of society (Ostrom, 1990, p.51). Governmental systems themselves are built upon institutions, such as the rules of intergovernmental relations, the principles of taxation, representative democracy, or rules of bureaucracy. Many of the institutions that should be considered relevant for the relation between environment and society are not only partly determined by government intervention, but they are even part of the (sub-) political system itself. The essential issue here is, that in so far institutions are product of state interventions, their design is mostly determined by the State's structuring, achieving functions and above all its steering function. Policy measures directed at the improvements of environmental management have to fit in with the institutions for economic development, which throws up several barriers for implementation. For example, the introduction of particular *pigovian* taxes (according to the principles of Pigou; section 5.2.1) in the European Union, for internalization of external costs, or particular packaging rules in member states, cannot be implemented without changing existing EU rules directed at free market development.

9.2.4 The arbitrating function

The arbitrating function is directly related to handling environmental issues. In densely populated areas economic growth and population growth resulted in a situation in which the supply of space available for economic activities was gradually exceeded by the demand. The arbitrating function may be compared to the function of an authoritative body in social-dilemma situations. That scarcity of space was later followed by 'new scarcity' for many environmental resources, so for that scarcity the arbitrating function is also needed. In this function the State has to act as umpire in conflicts arising from the new scarcity of environmental resources.

When the solution of environmental problems is seen to be the responsibility of the State, this arbitrating function must be reinforced. The use of environmental resources and the utilization of pollution sinks is a matter of division of scarcity, so it will be about conflicting interests. In the words of Mill (section 9.2): 'What rights, and under what conditions, a person shall be allowed to exercise over any portion of this common inheritance cannot be left undecided.' In our society the use of the environment is free, in principle. The institution of 'hardly restricted use' of environmental functions has to be changed by State intervention, as the quotation from Mill implies. The problem here is that such arbitrating decisions are taken by government bodies which have themselves important structuring, achieving and steering tasks, which are primarily exercised from the perspective of economic development in a competitive setting.

If a decision must be taken with respect to spatial arrangement, as in the case of zoning plans, then competing uses must be weighed against each other: agrarian vs. industrial, transport vs. housing, the landscape vs. the extraction of natural resources. The State must not only mediate between the conflicting interests of its citizens, it must also contribute to the welfare of all its citizens by creating the necessary conditions for economic activity, employment, etc. In this way State arbitration is coloured by the objectives of the decision-making body itself. While this may be perfectly legitimate, it does nothing to make the role of the State more credible, especially when it comes to long-term environmental effects. And in some cases it is not even legitimate. There are countless examples of public bodies which are themselves involved in polluting activities. This is certainly not only the case in former planned economy states in Eastern Europe. It is true in all western pluralistic capitalist states as well. The standard-setting, monitoring, and maintenance tasks are disregarded, because the consequences for the economy carry greater weight. In the past many nations, provinces, counties and municipalities have been involved in illegal dumping of hazardous waste, because they were too quick to issue permits, failed to enforce regulations, or were happy to export the impact of these hazardous substances (Wynne, 1987).

The structuring function of the State has taken shape primarily through general, sustaining regulations pertaining to the legal relationship between one citizen and another, and between citizens and the State. This function is guaranteed, because the responsibility for carrying out these regulations has been entrusted to the independent bodies of the judiciary. The authority of the State can only be extended to deal effectively with environmental problems when environmental values are assessed independently of other, notably economic values. At the moment this arbiter function cannot be guaranteed by modern States, because environmental regulations are not enforced by a power which is independent of the achieving and steering functions.

In some countries bodies with a measure of independance are created with limited power, such as the Environmental Protection Agency in the USA. Generally speaking such organizations do not exist at a power level that is creating an independant arbitrating function. The arbitrary function is effective only in cases where the judiciary can be used to enforce environmental values. If environmental values are embodied in legislation that is offering opportunities for civil law procedures, courts may be effectively used in making environmental values prevail. Legislation on liability, for example, may be used effectively in cases of environmental damage.

Development of public bodies which are as independant as possible appears to be crucial for effective environmental policy. According to Jänicke (1990, p.223) two factors determine that effectiveness: a cultural and an institutional. Jänicke states that the political culture should be one of openness with a strong capacity for consensus building. The institutional factor contains openness of the judiciary, a cognitive system open to information which diverges from usual patterns, an economy that is open to new developments, and last but least responsiveness towards renewal of the political system. In yet another study, in which English and German policies on acidification were compared, it was concluded that the German political system and political culture allowed a much earlier recognition of the acidification problems as well as a better policy development. In particular centralization in the UK is greater, whereas the division between German Federal and *Länder* governments made the political system more open to the new issue. This division has encouraged competition which provided momentum for the improvements of environmental protection (Boehmer-Christiansen and Skea, 1991, p.114).

At present that openness is only partly guaranteed in some countries (Eder, 1996). Generally such activities as monitoring and publicity are often done by environmental organizations which are not state agencies. While they are sufficiently independent of political considerations, they do not have the power to lay down sanctions. What is needed for the arbitrating function is a new implementation, or even an extension of 'the separation of powers' (also section 3.2.3). This priciple was developed by Locke while he was defining the guarantees of the (property) rights of the citizens. The familiar 'trias politica' of Montesquieu (..-..) was the form of the separation of powers that was implemented in most western nations.

For an improvement of the arbitrating function, the task of control and enforcement should be entrusted to more or less independent public bodies, not only in the field of civil law but within public law as well. Such a model already exists for financial policy in most western states, whereas national banks are public bodies that can operate in cooperation with the government, but structurally fairly independant from it. Another type of independant judgment over governmental actions is offered by international courts of justice. Although in most cases these have no jurisdiction in the field of environmental issues, in some cases, such as the European Court of Justice, there are possibilities to act independantly from national governments. There are even possibilities to prosecute them (example p.196).

The instances in which the authorities have been actively involved in the dumping of toxic wastes are a clear indication that the control and enforcement of environmental measures are deemed less important than economic interests. What is needed is not more powerful government, with far-reaching authority, but rather more restricted government. This will not be accomplished by limiting the tasks

Brussels puts Ireland in the dock over eco-damage

Ireland is being prosecuted for failing to safeguard sensitive habitats and unique landscapes, after conservationists complained that peat extraction and forestry had been allowed is several areas before environmental impact assessments were conducted. The government will face charges at a hearing in the European Court of Justice later this year that at least 256 hectares of The Burren's distinctive limestone paving and alpine vegetation, in Co Clare, have been destroyed in the past two years. (...) A further charge is that European funds were used simultaneously both to conserve and develop Clonfinane bog - a unique peatland in Co Tipperary designated of international importance. The office of Public Works in Dublin spent Pounds 300,000 of European money buying up part of the 800 hectare bog at the same time as it received funding from Brussels to develop a conservation programme for the peatland.

Sunday Times, January 12, 1997.

and responsibilities of the authorities, but rather by redistributing the power of the State. Owing to the recession of the early eighties, there has been no call for an extension of the powers of the State. In fact, in many countries a discussion has arisen on governmental withdrawal and 'deregulation', often prompted by the budgetary problems of the central government, and the slowdown in economic development. There are references to the 'stagnating welfare state' and the 'crisis in the welfare state', reflecting doubts about the quality of government performance. International comparison of the effectiveness of environmental policy shows that there is no 'withering away of the state'. The regulatory mode of intervention by public administrations has shown to be crucial in almost any case of improvement of environmental quality (Jänicke and Weidner, 1995, p.21). In the countries of the former Comecon, the withdrawal of the government presents problems of an entirely different nature. The state monopoly and the absolute control over all aspects of society has had a disastrous influence on almost all forms of economic activity (section 3.2.2). At the same time the environment was damaged in many ways. For that reason it may perhaps be helpful to define what should and should not be included among the powers of the State. It is important to point out to these countries that in Western Europe (limited) state intervention and particularly clear regulation have been the driving forces behind economic progress, and that in countries with a high degree of *laissez faire,* in which the government has almost fully withdrawn from the economy, one of the first victims is the environment.

9.3 Unequal distribution

Environmental goods are in many ways vital for society, but in economic terms there is no free market on which prices or cost are assigned to them.

Environmental effects constitute external costs and because of the social dilemmas involved, limiting these side effects requires a collective approach (section 5.3). Whether such an approach is actually undertaken by the State will depend primarily on who is actually affected by these external effects and the forces which they are able to mobilize.

9.3.1 Peripherilization

Although environmental goods are economically described as collective goods, suggesting there is utility for everyone connected to them, in practice the burden of damage to environmental goods is by no means evenly distributed. This implies these environmental goods are not of equal value to all. Any action taken to reduce the impact of damage to environmental goods or to prevent that damage is of particular value to the one that is suffering from it.

An example of this unequal distribution, in which power is a determining factor, is the phenomenon of *peripherilization*, the clustered and dispersed siting of hazardous facilities (Blowers and Leroy, 1994). In general, communities will be peripheral when their socio-economic structure is weak because they are dependent on a single industry. This phenomenon is established repeatedly in connection with nuclear facilities.

An example of that weakness and a reinforcement of it occurs when decisions have to be taken on the site on which nuclear waste should be disposed of. Blowers *et al.* (1991) describe the efforts of the British Nuclear Industry Radioactive Waste Executive (NIREX) to find a site for underground radioactive waste storage. After many efforts elsewhere, which were blocked by opponents in the political process, the only suitable place left was what they call a 'nuclear oasis', Sellafield, a location which already had nuclear installations. That site was not chosen for it geo-scientifically established suitability for nuclear disposal, but because of the fact that it was expected that the opposition there would not be able to effectively resist the siting owing to the political and economical weakness of the region, particularly the already existing dependancy on nuclear facilities. Neverheless, even that appeared to be a mistake, as governmental approval for the disposal has been refused in 1997.

When new nuclear plants or nuclear waste deposits are planned at existing sites there is in general less opposition than when they are built at new sites and the dependence of these areas on the nuclear industry is reinforced by such outcomes. 'Clustered nuclear siting has been followed particularly in Canada, Belgium, the UK, Finland, and Sweden. In all these countries, local opposition was largely avoided by such siting approach. Countries with a highly dispersed form of siting, such as the USA and Germany, experienced a far higher incidence of strong local opposition' (Rüdig, 1990, p.298).

When nuclear facilities are not clustered, they still tend to be located in economical weak areas. Slovic et.al (1991) established in the United States that

Waste makes haste

Nuclear power has not turned out to be the blessing the advance men said it would. Among much else, they presented it as clean - no more burning of gritty coal - but in the matter of cleanliness, it has a ghastly problem of its own. The nuclear issue is waste disposal - what to do with the enormously toxic spent fuel rods for which there currently is no long-term home. (...) In the 1980s Congress fastened on Yucca Mountain in Nevada as a likely permanent repository. Nevadans resisted the idea, but Texas and Washington, the other candidates, were more powerfully represented in the House and able to duck. (...)

The Washington Post, July 16, 1996.

Senate passes bill to create nuclear waste dump.

(...) Taking a step toward resolving the long-running controversy over where to store thousands of tons of nuclear waste, the Senate passed a bill Wednesday to create an interim storage site at Yucca Mountain, northwest of Las Vegas. (...)
Nevada's two Democratic senators had managed to tie the measure up with a series of parliamentary manoeuvres. But on the House side, the state's tiny two-member delegation seems unlikely to be able to muster similar opposition. 'It is a travesty,' said Nevada Sen. Richard H. Bryan. 'This is special-interest legislation at its worst .... with big bucks behind it.' But the Nevadans faced a formidable dilemma: Their vociferous opposition to the bill has been accompanied by sighs of relief from other states that are not being considered for the waste storage facility.

Los Angeles Times, August 1, 1996.

regions may also become nuclear dependent. There was a reinforcement of the dependence on these activities in the case of the establishment of nuclear facilities in Nevada. An investigation among managing directors of other branches of industry showed that they were less willing to invest in Nevada as a result of building these facilities. Stronger regions can resist hazardous activities more easily, and they are likely to do so when they recognize the decreased willingness to invest in their area.

These kind of issues are examples of political decision making on the distribution of hazards within the Risk Society (section 7.7.3). The lack of opportunities for external distribution of hazards will eventually grow to the primary, central object of politics (Beck, 1992, p.183).

This peripherilization not only occurs on the regional level, but there are issues of distribution of hazards on a global scale and the local scale as well. An example of social consequences of distribution of hazards is given by Greenberg and Schneider (1996) who investigated environmental hazards in relation with the quality of neighbourhoods. Although generally environmental concern among blacks averagely is lower than among whites (Taylor, 1989), Greenberg and Schneider found more concern about hazards among black people than among whites. It is a phenomenon which is linked with relatively large number of blacks living in neighbourhoods with hazards, as established by the US Environmental Protection Agency (EPA, 1992).

9.3.2 The environment: a problem for the happy few?

There are a number of tasks which the State can carry out. As we have seen, during and after the industrial revolution economic development gradually became the primary task of the State. Any new tasks which are taken on will have to be weighted against those economic interests. This was true of social services, and it is equally true of all the tasks related to the environment.

According to De Swaan (1987), the probability that in a community a collective process will be initiated to restrict environmental effects, increases when a greater part of it is borne by the elite of that community. In that collective process State agencies or other collective arrangements may be called in to deal effectively with the environmental threats. In other words, State measures to deal with environmental damage are taken in those cases where it is mainly the elite that is affected. The question is then who is in a position to make use of the State and its mechanisms to further her or his own interests? In short, whether or not the State is taking effective measures is not primarily a matter of rational decision making, it is also matter of power.

To answer the question of when an environmental problem becomes a task for the State, we must examine not only the social groups who bear the burden of the environmental effects, but also the groups involved in causing and solving the problem. Ultimately it is a matter of interests, among which we can distinguish victim interests, polluter interests, and support interests (Von Prittwitz, 1990). These three categories may also be distinguished by answering the questions:

- ★ Who has an interest in solving the problem?
- ★ Who benefits from the interventions which cause an environmental problem?
- ★ Who benefits from involvement in the actions undertaken to solve the problem?

Generally speaking the probability of effective environmental policy decisions grows when victim interests can be found within more powerful groups within society. That probability will also be greater when support interest can be found among powerful actors, while the probability will be low when polluter interests are concentrated in particularly powerful actors who do not have an interest in the problem solving measures and who do not have a vicim interest. As a result of the separation between the advantages and the disadvantages of activities affecting the environment, the solution to environmental problems invariably involves a weighing of interests. This leads to conflicting interests and ultimately to actual conflicts. Whether or not State measures are used on behalf of the environment will depend on balances of power, wich will be discussed in the following sections.

9.4 The political process

At several different points in the present discussion we have maintained that when there is a separation between the advantages and disadvantages of activities which have environmental consequences, this leads to conflicts of interest. In this situation, it is the balance of power which determines whether the State is on the side of the environment, or is used to promote other interests.
The process of interactions by which binding values are allocated authoritaritatively for a society is known as the political process (Easton, 1965, p.21). 'Allocating values' means deciding on matters which are of importance for the entire society. Although for the most part this laying down of binding values takes the form of government decisions, the definition of the political process is not confined to what is generally referred to as 'politics'. Although the ultimate decision on some social problem may be taken by a government body, many different groupings have played a role in the run-up to that decision. Companies, churches, associations, standards organizations, trade unions, lobbies, and action groups are all 'actors' in the political process.

9.4.1 Power

Policy is mostly described in terms of a rational decision making process, in particular in policy documents. Problems are detected and analysed, options for solving these problems are formulated, goals are set and the tools to achieve these goals are selected. However, in all stages of the process of policy formation decisions have to be taken within the framework of politics. The political process follows ways other than the one directly from problem via setting goals and selecting instruments to the solution. The political process is far from that rational one suggested in policy documents, not because actors involved act irrationally, but they all act according to their own objectives which are largely determined by conficting interests. So all actors are biased, and as they do not control the process their actions are not only based on their objectives, but on the assessment of the ways other actors in the process will behave. Their involvement in the process is partly determined by strategic considerations. This is effectively summarized by Schattschneider in a famous quotation on the process of democratic decision making in America: 'All forms of political organization have a bias in favor of the exploitation of some kinds of conflicts and the suppression of others because organization is the mobilization of bias. Some issues are organized into politics while others are organized out.' (Schattschneider, 1960, p.71).
A crucial concept in the political process is power: who is able to 'organize' issues into or out of politics. An extreme example of it is given on p.201. Power is the ability of actors to lay down alternative modes of conduct for other actors. Within political science there is an understandable interest in such concepts as 'power' and 'influence'. In the course of many centuries, a variety of views has

Endangered species convention.
Japanese 'bribes' threaten wildlife.

Japan has been accused of 'buying' the votes of small Caribbean states with overseas aid in order to block efforts to save endangered species such as elephants, turtles and whales. A report on Japanese 'bribes' is being released to delegates at the Convention on International Trade in Endangered Species (Cites), which opens in the Zimbabwean capital Harare on Monday. (....) The 'vote consolidation operation', as Tokyo's fisheries' ministry documents describe Caribbean aid, is designed to defend Japan's interests against the environment lobby. Among the examples of aid quoted are grants totalling pounds 19 million to St Lucia between 1987 and 1995, and a similar amount to St Vincent over the same period. (...)
Japan's 'vote consolidation programme' was first suspected in 1992. A bloc vote at the International Whaling Commission, shored up by five Caribbean islands siding with Japan and Norway, has often been sufficient to hold off adverse votes on whaling.

The Guardian, June 7, 1997.

been voiced, and gradually there arose a need to measure the concept of power, in order to determine who has power, where it is concentrated, and how. It was the various studies on power carried out since the fifties which triggered the further development of this concept.

9.4.2 'Non-decisions'

In a series of studies aimed at tracing the origins of power, various methods were developed to measure power. One of these is the so-called reputation method, whereby individuals who are part of the social system under study are asked who they consider to be influential or powerful. Another is the participation method, which focuses on the activities of individuals within organizations, while other methods consider the position of the actors to be most important.
In the decision method, a number of selected problems are put forward in order to establish which actors have tried to exercise influence, in which areas, at which point, and by means of which activities (Dahl, 1961). The use of this method has greatly influenced our thinking about power. The decision method deals with decisions which have actually been taken, and does not take into account the fact that an actor may be more involved in other ways of 'mobilizing bias' (in terms of Schattschneider). Powerful actors are primarily concerned about sustaining a given mobilization of bias by means of 'non-decision-making', as Bachrach and Baratz (1970, p.43) name it. If actors are able to prevent certain decisions from being taken, then this also constitutes a form of power. Bachrach and Baratz refer to this as 'the other face of power', while the term non-decision was used to describe the phenomenon.
Non-decision-making was illustrated in a comparative study of how the problem

of air pollution was dealt with in two Chicago suburbs (Crenson, 1971). Crenson investigated the extent to which the subject of air pollution was an issue on the political agenda and whether this ultimately led to government measures. His conclusion was that the power of industry had kept air pollution from becoming a topic of discussion, not by influencing the contents of the decisions, but by determining whether the problem was even discussed in the political arena.

9.4.3 Environmental issues on the agenda

The change of existing behavioural patterns is fundamental to all environmental policy. Alas, 'business as usual' often has strong advocates involved in sustaining the given mobilization of bias, that is vested interests that strategically will opt for nondecisionmaking. Often the introduction of environmental goals in policy concerns new lines of policy added to existing policies in several policy domains. The penetration of new policy principles and goals in historically grown policy domains is always an uphill battle, within all policy systems at all levels, in either national or local political arenas.

For substantial decisions resulting from changes in current policy very specific conditions have to exist at the same time. Information about environmental problems, and knowledge about the ways to prevent or solve them, will never be sufficient. Effective new policy will only start at moments when different social developments and events come together. In the agenda theory of Kingdon (1995) it is not even a problem which can reach the stage of the political agenda, it is rather a problem-solution pair. Agenda status will only be achieved under certain conditions, including the condition that a mutually attuned solution-problem is already available. Only then is a so called 'policy window' created. Environmental issues first have to appear on the political agenda, which in many cases does not happen, as Crenson illustrated for local air pollution issues. For an explanation of this phenomenon we will look at the structure of policy subsystems. These tend to be dominated by a coalition of social and political actors, which is a structural barrier for policy oriented learning, the development and implementation of new policy lines. We will first discuss the theoretical framework with regards to these coalitions, which are called 'advocacy coalitions'. Then we will outline a model of the political proces, in which the concept of non-decisions is elaborated.

9.4.4 Advocacy coalitions

A theoretical approach to explaining the stubborn resistance to changes in policy is the so called advocacy coalition framework (ACF) which 'argues that policy change is best seen as fluctuations in the dominant belief systems within a given policy area/subsystem over time. While policy analysis and learning can strongly affect secondary aspects of such belief systems, fundamental changes in subsystem policy are usually the result of alterations in noncognitive, systematic parameters.' (Sabatier, 1987, p.650). The framework is a model applicable to the evaluation of policy changes, which includes all actors and institutions involved in a *policy subsystem*. It has been used in several cases of development of environmental policy, such as nuclear waste disposal (Jenkins-Smith, 1991), car pollution control (Grant, 1995), waste management (Eberg, 1997), regional environmental policy etc. (Sabatier and Jenkins-Smith, 1993).

A premise in the ACF is that the unit of analysis for studying policy change should not be any governmental organization, but rather the policy subsystem, which is the set of actors within a policy domain. Actors at all different governmental levels participate, as well as many actors who are not part of the State. When public as well as private actors who have shared policy beliefs are cooperating they form an *advocacy coalition*. Within a subsystem competing coalitions may exist and by definition these coalitions hold different policy beliefs. For example: in the case of waste policy in Bavaria Eberg (1997, p.84) traced two coalitions. The dominant Incineration Prominent Coalition consisted of the Environmental Ministry, the private and municipal waste incinerators, the waste management firms and the energy utilities. The Incineration Prevention Coalition was composed of environmental organizations and the environmental political parties. Generally there is a dominant advocacy coalition, which means that existing policy is based on one dominant *policy belief system* (PBS) which contains three levels:

* The *'deep core'* that contains basic normative assumptions, that are not specific for one policy domain. An example of such a deep core belief is the ideology that all citizens are primarily driven by self interest (also see section 5.8) and that all policy should primarily be determined by that.
* The second level is the *'policy core'*, containing the general strategy for achieving the assumptions of the deep core beliefs within the policy domain. In this second level fundamental assumptions about the policy domain itself are also represented. Environmental issues tend to come up within subsystems that are primarily concerned with economic development, so economic values will mostly be dominantly present in belief systems.
* The third level of the PBS is a set of *secondary aspects*, which are mainly about instrumentation and the information needed for implementation of the 'policy core' within the policy domain. That instrumentation also encloses the rhetorics, for example the rational justifications in policy documents.

The policy beliefs are about matters concerning the particular features of a policy domain. Policies always contain assumptions about the way society will react upon policy interventions. These assumptions include implicit (although in some cases they are made explicit) theories about the best ways the objective of a policy can be achieved.

The three levels of the PBS represent a decreasing order of resistance to change. The central thesis is that cognitive factors will not be sufficient for policy oriented learning, except concerning secondary aspects of policy beliefs. To put it bluntly, official public policy subsystems hardly 'learn' in a cognitive sense. Real 'learning' followed by new policy, needs a change of policy beliefs, at least at the policy core level, and sometimes even of deep core beliefs. For fundamental changes in policy the existence of competing advocacy coalitions is crucial, as such changes may only occur as a result of another coalition becoming the dominant one, replacing the old PBS by its own. This is not going to happen easily, and it certainly does not happen solely because of new information and knowledge. Events external to the policy subsystem should help to achieve changes within the subsystem. These events include changes in socio-economic conditions, changes in the governing coalition, policy decisions and impact from other policy domains and changes in public opinion (Sabatier and Jenkins-Smith, 1993, p.224). Actors within the subsystem will always try to incorporate these changes within existing policy. This can be done by adaptations in policy beliefs, however, only by a change of secondary aspects. Changes in policy in reaction to external events are directed at defending core and policy core beliefs.

Many efforts towards changing policy into new more sustainable directions meet barriers in the process of policy oriented learning. For the implementation of environmental policy core beliefs in many policy domains should be changed. Environmental policy in many cases means a change within domains that already have long existing policies, based on strong belief systems and relatively stable advocacy coalitions. As a result environmental policy will mainly not be more than a change in secondary aspects of existing policy. For example: a policy directed at gaining control over motorized mobility because of environmental impact (air pollution, carbon-dioxide, congestion, exhaustion) will have to fit within the policy and the PBS of the dominating advocacy coalition on transport. Concrete measures will only be taken if they concern secondary aspects of transport policy.

9.4.5 Barriers in the process of agenda setting

Non-decisions come about in two ways: by keeping an issue from being put on the political agenda (prevention of issue formation), and by interfering with the resolution of a problem that has reached the political agenda (demand perversion; Van der Eijk and Kok, 1975, p.283). In other words, social problems on the way to solution can be tampered with as well as delayed.

On the basis of these considerations a so-called barrier model for the political process has been developed (Bachrach and Baratz, 1970; Cobb and Elder, 1971). In these barrier models the progress of a social problem on its way to the political agenda and then to political decision-making is described in terms of phases. In the literature this model has become known as the agenda approach to the political process and the environment is an obvious field in which to employ this approach. We distinguish four phases in the barrier model of the political process:

I Social *needs* or wants: a problem within society which has been recognized, and for which a solution must be found; for example the recognition by Molina and Rowland (1974) of CFCs damaging the ozone layer.

II *Demand*: a problem for which one or more groupings are demanding a binding solution; the demands by some environmental groups for discarding CFC-production of propellants and cooling.

III *Public issue*: a demand or want which is already on the public agenda, and thus has public support; the public asking for spray-cans without CFC propellants.

IV *Political item*: a demand which has made its way onto the political agenda, and on which debate is going on in the run-up to a decision; several national limitations (US: 1978; EU: 1980) on CFC-production and specific uses, later followed by the Montreal treaty in 1988 (Benedick, 1991a).

Even when the stage of a political decision has been reached, there is aways a possibility that the decision will ultimately not be implemented. Political output may be symbolic, or it may be perverted.

The analytic distinction between these four stages is based largely on the extent to which public opinion recognizes the existence of the problem. The four phases are analytical, and not necessarily chronological. The transitions between them are marked by a number of empirically demonstrable barriers. The barriers in the political process are mainly the result of the structural relations between the actors involved. These structures particularly prevent policy oriented learning based on merely cognitive factors, as discussed in the previous section. Barriers arise out of the social structure of all organizations and groups which are involved in dealing with the problem, their position within society and the relationship between them. More specifically, barriers arise as a result of the political structure, including all governmental and semi-governmental bodies. Finally the nature of the problem determines the barriers to effective policy measures. In the case of environmental problems, for example, it is important to note that these are by definition the result of human intervention, and therefore can always be traced to social causes.

At each of the identified four stages in the political process the process may stagnate. Where environmental problems are concerned, demands are often formulated by environmental organizations, whose work consists largely in lobbying public opinion, because they are seldomly part of the dominant advocacy coalition. They are not always successful in getting a particular issue onto the political agenda, and this illustrates the first barrier to obtaining a solution: the problem is simply not put before the public. It is not only a question of how much people know about changes taking place in the physical world around us, but above all the recognition that these changes are harmful to the environment. In countries where the primary task of the State is the creation of optimum conditions for economic development, there are very few bodies geared to recognizing environmental problems, although obviously the number of public bodies such as the Environmental Protection Agency in the USA, has increased significantly in many countries. An intervention in the structure of policy subsystems which can help to identify environmental problems is the creation of independent bodies, such as advisory councils, which can take the initiative in getting these problems onto the agenda. The impartiality of such bodies is crucial.

9.4.6 Redefinition of problems

The existence of barriers can also mean that in the course of the transition from one stage to another the problem and proposed solutions are revised, redefined, or 'perverted'. In the course towards the political agenda, the process of redefinition is causing substantial bias. Many environmental issues may be viewed as social constructions (Hannigan, 1995). This does not mean the environmental problems are 'invented' and the physical features that constitute the problem are mere fiction and do not exist in reality. It means the way these problems are defined and the significance that is attributed to the supposed physical and social consequences, is only partly determined by the physical reality, as far as this reality is known with a fair degree of certainty in the first place. Scientific fact finding is also part of this process of social construction. As Jasanoff describes in an analyses of the way the US Environmental Protection Agency operates, scientific fact finding processes accommodated themselves to the demand of politics (Jasanoff, 1992, p.23).

The manifestation of the issue is partly a social construct as well (see also section 10.7.1). An example is the damage to the ozone layer caused by chlorofluorocarbons (CFCs). As Benedick (1991) claims, the scientific community was the driving force behind the ozone layer becoming a policy issue. At the time the problem was identified in 1974, these CFCs were found mainly as propellants in spray cans. In the long term, however, the major threat has come from vaporisers for foam plastic, and above all from the CFCs released by discarded refrigerators and car air conditioning units. Not only do the CFCs damage the ozone layer, they also contribute to global warming, in particular the present prime alternatives to CFCs, the HCFCs (Kroeze and Reijnders, 1992).

Environmental groups in most countries saw government policies as much too lenient. One reason for this was the fact that the problem was defined differently by governments and the environmental groups. The latter saw the rejection of pollution as part of a rejection of wasted energy and non-essential 'luxury' products. In their view, the narrowing of the 'CFC issue' to a 'spray can propellant issue' represented a perversion of the issue. The way it was defined appeared to be important for the problem to become a public issue. Within international politics, even the way the scientific community has promoted the issue for taking effective policy measures, has also been subject to construction (Yearley, 1996, p.110).

9.4.7 Decisions which do not solve anything

Not every decision at the political level leads to a 'solution' to the problem in question. In many cases symbolic decisions are taken. In many other policy measures are being propagated even before it is clear which problems have to be solved, a phenomenon Kingdon (1995) refers to as 'solutions looking for a problem' (also section 9.4.3). A nice illustration of this phenomenon can be found in the struggle about the implementation of nuclear power in many countries. These struggles mainly concentrate on the safety issues of nuclear reactors and the storage of nuclear waste. Although in particular the waste issue is not solved anywhere, many decisions about storage have been taken in all countries using nuclear power. Many environmental problems are used as an argument in favour of nuclear power, among them global warming (Pasztor, 1991). Although the actors and coalitions advocating nuclear power are doing so for many decades now, carbon dioxide reduction now seems to be their main argument although they became aware of this issue only a decade ago.

Even if political decisions on environmental problems are ultimately taken, that does not necessarily mean that they will be beneficial to the environment. The safety problem represented by nuclear power plants is an extreme example of the way an item can be removed from the political agenda. In most cases the alternatives which benefit the environment are weighed against alternatives which are favourable to other interests. The outcome is determined not so much by the relative merits of those interests as by the power and influence of the groups which back them. This means that in some cases the outcome is paradoxical. A case in point is the Clean Air Act in the United States. A special coalition against the legislation was formed by industry, power companies, mining companies, and many others, who had changes made in the standards and the way they were to be implemented (Melnick, 1984). While the Clean Air Act was intended to reduce pollution, as a result of the amendments pollution actually increased in certain regions. The major negative consequences for the environment resulting from the amendments include an extension of the time allowed for air purification, and an increase in the space reserved for the dumping of mortar wastes, leading to large increases in the volume of those wastes (Navarro, 1981, p. 133).

Governments often display a strong preference for measures which are only minimally coercive, such as public information campaigns. Although it is an accepted fact that when employed on their own such campaigns are seldom a success, they are often the measure preferred by the authorities. By opting to provide the public with information in an effort to solve an environmental problem, the authorities avoid the necessity of a political decision that is based on a real choice. At the same time they appear publicly to tackle the problem at hand, although it is mere 'window-dressing'. In many cases the decision to carry out an general information campaign is a form of non-decision-making.

9.4.8 Pressure of public opinion

A crucial factor in the setting of the political agenda, is public opinion. In the four stage model of the political process the public issue stage has been distinguished. In general public opinion may have two kinds of influence: as a push factor and as a necessary condition.

Public opinion can be the driving force putting an issue on the political agenda. A problem that has been developed to the stage of public issue cannot easily be avoided by the authorities any more. In many cases they may tend to do so, forced by strong interest groups in the society. In the theoretical framework of advocacy coalitions, changes in public opinion are a factor that, under certain conditions, may induce policy changes. However, it is described as a factor outside the policy subsystem (Sabatier and Jenkins-Smith, 1993, p.224). When public opinion strongly demands a solution for an issue, it will be an important factor that may no longer be ignored.

Environment-bashing could cost republicans at polls
by Gary Lee, Washington Post Service

A leading Republican pollster is warning party leaders that trying to cut spending on the environment and roll back environmental laws could cost Republicans dearly at the polls in November.

'Attacking the Environmental Protection Agency is a non-starter,' concluded a report by leading party pollster, Linda DiVall. 'Republicans should be emphasizing the safeguarding of reasonable and balanced environmental protection done in a more efficient manner.' (...) The survey was commissioned by the Superfund Reform Coalition, a business organization seeking to reduce the costs, imposed on industry by the Superfund law, of cleaning up toxic waste sites. (...) The survey found strong backing for federal laws that protect the environment. Forty-six percent of respondents said no changes should be made in the 1972 Clean Water Act, the blueprint for combating pollution in the nation's lakes and rivers. Respondents also expressed support for federal statutes protecting endangered species and supplies of drinking water.

International Herald Tribune, January 25, 1996.

Secondly, public opinion may be a crucial condition for policy changes. Many solutions to environmental problems that authorities may choose, can only be real solutions when the decisions are accepted by society and can be implemented effectively (see section 9.6.2). The possibilities for solving environmental problems will be much larger when the issue is recognized by the public as an important one.

The difference in the role of public opinion as a push factor and as a condition depends on the origin of initiative within society to put the issue on the agenda.

Cobb *et al.* (1976) distinguish between 'inside access', in which the initiative to deal with the problem lies within the official policy agencies, and 'outside access', with the initiative somewhere else in the society. For outside access public opinion is an instrument for getting policy response. In many cases public opinion is an instrument for environmental groups such as, for example, for Greenpeace. Public opinion, however, may be a powerful instrument, but it is not easy to handle. Mass communication is not directly within the hands of environmental groups. Media, such as television and newspapers, play an important role in getting environmental issues on the political agenda, but analysis of media content shows that official views as issued by representatives of authorities have significantly greater access than the views of local critical and environmental groups (Stallings, 1990).

As noted earlier, in the course of the transition from one stage to another the problem and proposed solutions may be revised or perverted. Several actors in the political process are trying to manipulate public opinion one way or another. An example of it can be found in the issue of disposing of old oil platforms, that are put on the political agenda by means of the Brent Spar case by Greenpeace (Dickson and McCulloch, 1996). In that case plans existed to sink it in a submarine valley in the Atlantic. The way Greenpeace tried to influence public opinion and finally the governments of European Union member states ultimately lead to spreading misinformation about the amount of toxic substances in the tanks of the oil storage facility.

Specific environmental problems do not reach the stage of public issue very often. Public opinion seldom concentrates on very specific political decisions. The attention mostly goes to large, broad defined general issues, such as 'criminality', 'unemployment' or 'the environment'. Nevertheless a large support for the general idea that environmental problems are important and that certain kinds of solutions are needed, will be a good condition for an environmental policy with decisions that really help to solve problems. In general, the climate of public opinion in which significant environmental policy decisions have to be taken is important. The general support for environmental policy is due to an attention cycle (Downs, 1972) and not at a constant level, a subject discussed in section 5.6.1.

9.5 A rational model for environmental decision making

The problem of getting environmental issues on to the political agenda has been recognized, and efforts are made to deal with it. In 1969 the United States passed a law stipulating that an extensive Environmental Impact Statement (EIS) had to be drawn up before particular interventions were permitted. In the past decades legal frameworks have been created in almost all Western countries for the reporting of the effects of serious environmental interventions. As a result of the introduction of such environmental effect reports, an interdisciplinary field of research centring on the assessment of environmental effects has been created, known as Environmental Impact Assessment (EIA). In principle EIA involves both the physical environment, containing abiotic, biotic and ecological components, as well as the social environment, containing personal, interpersonal and institutional components (Erickson, 1994, p.6). A narrow interpretation only includes the physical environment, while for social effects the Social Impact Assessment (SIA) was developed. These instruments, which are based on legal obligations to report environmental impact, were primarily meant for the recognition of environmental effects from interventions like the siting of facilities. In that way environmental issues would be put on the agenda. The result of the introduction of mandatory impact reports has been the development of procedures of rational decision making, in which theoretically environmental impact and its social consequences is weighed against the benefits.

9.5.1 Impact assessment

While in the United States social effects are covered by EIA they are not included in the official impact assessment procedures of most European countries. The tendency to disregard social effects can be traced to opposition from those who take the initiative in projects likely to be accompanied by environmental effects. In their view, 'social effects are always about costs, and never about benefits, which is why they are always used to halt or inhibit developments' (Burdge, 1987, p.143).

Many topics may be investigated in the course of SIA-research. Topics concerning the way people appreciate their surroundings, such as changes in values, beliefs and attitudes; and developments in community life, such as the impact on health and human services, triggered by environmental changes. Another category involves the community resources, such as developments in energy supply and demand and water resources, but impact on natural resources like wildlife, forests, scenery and parks as well. Social impact of environmental changes may also be important for the economic development of communities, as impact may be traced to available manpower and level of education, on industrial and recreational development, or on things like mass transport.

9.5.2 Rational procedure to decide

The reporting of environmental and social effects should be part of the decision making process on any future project. In most procedures officially established in many countries, background research and the assessment of the ultimate effects are only steps along the way. However, just as important is the manner in which the decisions are taken, and the role of those affected by those decisions, not least the local population.

A complete impact assessment process theoretically includes ten steps (Wolf, 1983):

(1) Scoping: assessing the size of the problem (area boundaries, time horizon, etc.)
(2) Identification: defining the problem and analysing the causes.
(3) Alternatives: formulating reasonable alternatives corresponding to identified concerns.
(4) Profiling: defining impact categories, criteria, indicators.
(5) Projection: performing trend impact analysis.
(6) Assessment: sensitivity analysis for alternative outcomes of alternative plans.
(7) Evaluation: reformulating criteria, ranks and weight preferences, trade-off analysis and identification of preferred alternative.
(8) Mitigation: identification of unavoidable impact, and identification of mitigation measures.
(9) Monitoring: measuring of actual changes vs. predicted changes.
(10) Management: defining a management plan for reducing impact.

A significant feature of the rationality of this theoretical process is the formulation and development of several alternatives (step 3). As impact assessment is always a matter of decision making, the acceptability of impacts can only be established when different alternatives are taken into account. This is very important, because as has been pointed out in section 7.6.1, environmental impact such as risks and annoyance can never be 'acceptable' in a universal manner. Only options can be judged 'acceptable' in comparison to other options. Hence, impact assessment cannot be done in a one-option situation, as in a process of rational decision making several serious options have to be developed. It is essential that during the debate serious consideration is given to alternatives to plans which would affect the environment. It has gradually become clear that negotiation will always be necessary when it comes to the nature and the significance of environmental effects. In the long run, ignoring the conflicts which inevitably accompany projects with an environmental impact will only create problems for implementing policy decisions. In the United States, for instance, it has been shown that the public invariably continues to question a proposed waste incineration plant, as long as alternatives which entail waste reduction are not taken into consideration (Siskind and Susskind, 1989, p.319). These conclusions are put forward by researchers involved in the reporting of environmental effects. In practice, however, such insights are not yet widely accepted. As a rule the funds made available by the authorities are intended solely for background research and an assessment of the effects (Finsterbusch, 1985, p.199).

Although a number of countries have gradually acquired a fair body of experience in the reporting of environmental effects, it is doubtful whether this has actually led to an improvement in the way the environment is treated. Impact assessments are part of the overall tendency towards the rationalization in handling ecological issues, which is part of ecological modernization processes (Weale, 1992; Hajer, 1995). In many cases impact assessments have deepened our insights into the effects upon nature and the environment of a variety of interventions. But one may well ask whether this has had any real influence on the ultimate decisions taken. Whether sufficient account is taken of environmental effects will depend not on the data contained in reports, but rather on the priorities employed by decision makers. In other words, it is a question of political will to take into account the effect on the environment.

As noted in sections 9.4 practise is mostly stronger than the theory of rational decision making. From the policy oriented learning approach we know policy changes are generally forced to fit within existing dominant policy beliefs. So all cognitive contributions towards new policy originating from impact assessments, may even be interpreted within the framework of dominant policy beliefs.

Although we are not sure about the influence of environmental and social impact assessment on the ultimate decisions that are taken, the assessment itself has shown to be very useful. A great deal of knowledge about all kinds of impact is now available, thanks to much assessment research. According to the theory on policy oriented learning (Sabatier and Jenkins-Smith, 1993) cognitive factors of this kind will never be sufficient for policy change. Only secondary aspects of policy beliefs may be adapted (section 9.4.4).

9.6 Environmental policy

The State is not the type of umpire who is above the various parties (section 9.2.4). As one of the actors, or rather, as a number of different actors, it is involved in many different ways in the interests of society. For the moment priority is being given to economic development, and it would perhaps be naive to expect the government to take vigorous new measures aimed at safeguarding the environment. However, this does not take away the fact that governments do have environmental policies. Despite all the limitations of the political process, there are a great many environmental problems which do appear on the political agenda, albeit in a perverted form.

The working of society can be described in three categories of 'mechanisms' significant for designing policy measures: coercion, persuasion, and transactions (Lindblom, 1977). Policy instruments can be classified this way, as these instruments are efforts to achieve changes of behaviour by utilizing these mechanisms (WRR, 1992). Beside this threefold categorization we can make distinctions between one-sided and two-sided instruments, and between behaviour limiting and behaviour broadening instruments. For example, a financial levy on certain activities, such as a wire charge on electricity consumption, is a one-sided, behaviour limiting transaction instrument. We will first discuss the one-sided instruments of the categories coercion and persuasion.

9.6.1 Coercion

Coercion is basically a category which can only be executed by public law, that is, when we refer to legitimate coercion. Coercive instruments are based on a one way relationship: the State is forcing people to act in a specific way, or it is restricting activities. Coercion involves all kinds of direct regulation, such as banning orders, permits and commands. Absolute banning orders are being issued for several substances in many countries, for example particular pesticides and asbestos. Sometimes limited bans are issued for substances in a limited number of applications, or limited in time and region, as in the case of CFCs.

Emission standards hasty, industry says
More study needed, auto makers, refiners say

Auto makers and refinery operators are demanding that the country's environment ministers postpone a push for national standards to get cleaner cars running on cleaner fuels on Canada's roads.
The ministers begin a two-day meeting in Whitehorse in Monday. But both auto makers and refiners have objected to the speed with which an intergovernmental task force has put together an action plan for the meeting.(...)
Details of the recommendations are not known, but one early draft circulated within governments estimated that cleaning up fuels, particularly by reducing sulphur levels, could force refiners to sped about $1-billion in equipment changes, add a 4200-million a year in extra operating costs, possibly force the closing of uneconomic refineries and add more than 1 cent a litre to gas prices at the pump.

The Globe and Mail, October 21, 1995

Coercion also includes standards that are used as conditions for permits. Emission permits or building permits usually contain a list of conditions on things such as, for instance, the amount and quality of the emission of substances, or the time of day of noise emissions. Standards are not only used for activities, they can also be linked to products. For example, emission standards used for car exhaust fumes force the car industry to design and produce less polluting models. In the category commands we find instruments like obligations for deposit refund systems for selected products, for the application of environmental accounting systems and environmental reporting in businesses. An instrument that has already been discussed more fully in section 9.5 is environmental impact assessment, an instrument based on the coercion for making environmental impact reports. This is a special case because this obligation is particularly coercive for authorities themselves.

The other side of coercive policy instruments is that they require significant efforts of governments and other authorities for monitoring and enforcement. Monitoring can be hard to achieve, for example because of lack of technical and scientific infrastructure, and all the other problems connected with risk analysis. Remember the example described by Wynne (1987) of the impossibility of identifying and testing the toxic features of all substances that appear on the market (section 7.7.2). In practice, enforcement is hard to achieve as well, due to the fact that it will affect economic activities in most cases and it conflicts with priorities stemming from the steering functions of the state (9.2.3).

9.6.2 Persuasive instruments

The category of persuasive instruments involves all social regulation on a voluntary basis. Information campaigns and education programmes are directed at improvement of knowledge and awareness, while the creation of better conditions (facilitating) for environmentally sound options may also add to changes in behaviour.

Basically communicative instruments try to affect behaviour by means of changing attitudes. This process of persuasion is generally laborious, owing to the many barriers between communication and establishing a sustained behavioural pattern. As described in section 6.9 there are problems with the penetration of communications content and with the acceptance of that content. Even if the content is accepted this does not mean that the stage of retention, the actual adapting of new behaviour consistent with a newly developed attitude, followed by conservation of the new attitude-behaviour pattern, will be reached.

The instruments may be applied by authorities or by public or private organizations hired or obliged to use them. In the field of energy conservation for example, demand side management programmes (DSM; see section 8.4.2) include many communicative instruments, but also facilitating instruments (Nadel, 1992). These can be applied either by energy utilities or by agencies hired by the government.

Communicative and facilitating instruments do not have to be directed at individuals only, they can be applied to influence the behaviour of organizations as well. The introduction of cleaner production processes in industry and small enterprises, for example, is dependant on available information on techniques and organizational conditions (section 8.3). Energy conservation options, waste prevention techniques, application of monitoring systems, and methods for internal environmental management, are not generally known and neither are the costs and benefits of all these activities. The costs and benefits are of particular significance for business decisions and information supplied to industries and other firms cannot be effective when it is provided in general campaigns. It should contain assessments of the processes and needs of the enterprise that should implement new options for its production process (Van Berkel and Kortman, 1993). Government policy may facilitate the information programmes with an infrastructure and with financial support.

9.6.3 Instruments affecting private transactions

Two-way relations in which matters are exchanged (substances, rights, money) are transactions. Often governmental bodies are not involved as an actor, nevertheless there is an important category of policy instruments based on transactions. The conditions for transactions between actors in society will largely be determined by the state in private law and these conditions can be changed to achieve environmental goals. A significant example is the way liability is regulated. In cases in which the interests of private persons or organizations are damaged by environmental impact caused by others, private law is regulating whether or not a polluter has to pay for the damage. Widening the conditions for liability and expanding the possibilities to enforce the rights of the victims of environmental damage is a powerful instrument in preventing environmental impact. Liability must be considered an economic risk linked to all activities. The probability of costs following liability, or the insurance premium to prevent these costs, will be taken into consideration when private actors decide upon their activities. Insurance companies will also force their clients to take actions that will limit the risks of liability.

Pollution fall-out on directors

Company directors and senior executives may be personally liable if they were involved in decisions which led to their companies causing pollution, and objectors are likely to include them in litigation (...).
In addiction to civil hazards, directors can face criminal prosecution for 'environmental crimes' in the Environment Act 1995 and Control of Pollution Act 1974. For instance, it is an offence to infringe the code of practice on controlled waste, and a licence is needed for some sort of waste. Boards can be prosecuted for 'consent and connivance', which includes pollution when a director ought to have known what was going on and failed to stop it.

The Daily Telegraph, July 29, 1996.

Another important issue for environmental quality is the way property rights are regulated in private law. As environmental goods are generally collective goods, they are accessible without charge. However, this free access can be restricted by coercive measures such as permits or quotas, which can also have a transaction component. This is the instrument of tradeable pollution rights, or tradeable permits, which is partly based on the theory of Coase (1960).

When the victims of pollution own property rights to use a natural resource, for example clean air, it includes the right that this resource is not damaged by any polluter. If a market is created for these property rights they can be sold. Then the polluter and the one who is suffering from pollution can bargain over the externalities. The right not to be polluted can be sold to a polluter. As for most environmental goods there is no market for selling pollution rights by a victim to a polluter, so pollution rights or rights to use a common resource are mostly created by issuing permits. These permits can be made transferable with the objective creating a market which offers a solution to the problem of the distribution of pollution rights or quota (see also section 5.3.2).

A subject related to liability and transferable property rights is that of legal responsibility. As modern society is highly dependant on the division of labour, most products and services contain contributions from many different organizations. In its full life cycle a product is made from several components made by different companies with materials and resources delivered by many different industries. Then it is transported and distributed, consumed and discarded by waste disposers. All phases of the cycle may cause particular damage to the environment, but it is not obvious who is responsible for it. Who is responsible can be regulated by law: is it the consumer of the product, is it the producer, or the disposer? When environmental policy applies the concept of producer responsibility, the responsibility for the environmental impact in all phases of the life cycle is laid at the door of the producers. The costs of avoiding the environmental effects as well as the property rights of the use of environmental goods have to be re-allocated among all firms involved in the production process. For that re-allocation of these so called 'transaction costs' the tradability of these rights may be helpful (Williamson, 1979).

The fundamental objective of producer responsibility is to avoid environmental damage by implementing prevention. The idea is that when producers are made responsible for the impact of the disposal of their products at the waste stage, that is, when they have to pay for environmentally sound disposal or when they are liable for environmental damage, they will be concerned with that at the time of product design and production. In practice the working of instruments like this is not so obvious, as their succes is largely dependant on the structure of sector in which they are applied (De Jong and Wolsink, 1997; also section 8.4.2). The idea of product responsibility is applied in Germany, for example, where producers are officially responsible for the package of their products. Most efforts of the industry has not been directed at the implementation of prevention, but at investing in separate removal and disposal of package waste (section 9.7.3).

9.6.4 *Transactions between the State and private actors*

A second category of transaction instruments is formed by exchanges in which public bodies or governments are one of the actors. These transactions are regulated in public law and mostly they are of an economic nature. They include financial transfers: taxes and levies as behaviour limiting instruments, and subsidies as broadening instruments.

All taxes affect the behaviour of citizens and organizations, but an important category of environmental instruments is formed by taxes based on the principles of Pigou (1920). In economics the loss of welfare for other parties caused by economic activities is referred to as an external diseconomy or a negative externality. Environmental impact generally can be considered as external cost (section 5.2.1). To create a mechanism to internalize these external costs, a tax on the use of natural resources or the right to pollute may be introduced. This instrument is generally referred to as 'eco-tax' when the purpose is a change to the tax system in which taxes on income are replaced by taxes on primary resources such as energy (Von Weizsäcker and Jesinghaus, 1992). Other categories are earmarked charges (levies) directed at funding environmental measures, and regulatory charges used as an incentive for behavioural change in which the revenues are returned to the economy, such as special fuel taxes (WRR, 1992, p.37). For the moment, however, the elimination of environmentally negative incentives in existing tax systems, such as the abolition of several exemptions from taxation such as for airplane fuels, is more significant than the introduction of new taxes (Huckestein, 1996, p.398).

Another kind of transfer which involves governmental actors is the broad category of subsidies. This is basically an instrument which is not directed at restriction of activities but at broadening the spectrum of activities. These are mostly temporary measures designed to provoke investments in environmentally sound activities. Economic feasibility of cleaner production methods or renewable energy sources can be increased. The effectiveness of investment subsidies is often questioned, however, as the environmental effectiveness of these new activities is not primarily determined by the mere existence of environmentally sound systems, but largely by the way they are operated. For example, regulations of minimum prices to be paid for electricity generated by renewable energy techniques such as wind power, appeared to be helpful in optimizing energy yield. They have proven to be more effective than investment subsidies (Wolsink, 1996). The actual investments in installed wind turbine capacity appear to be larger in particular. European countries like Denmark and Germany that introduced reasonable prices for wind generated electricity by regulating the prices of energy sold to and bought from energy utilities, were quite successful in their implementation of wind power compared to countries that used investment subsidies.

A relatively new instrument is the policy covenant, which is a voluntary

agreement between a public actor and a group of private actors, such as an industrial sector. Although in principle these agreements may contain paragraphs of a legally binding nature, in practice legal enforceability is not the main concern of these covenants (WRR, 1992, p.48). The primary objective is to achieve environmental goals through consultation and voluntary co-operation. The idea behind the use of covenants is that the acceptability of voluntary agreements is high, and acceptability is generally a problem for many policy instruments. The problem of acceptance will be discussed in section 9.7.
Firstly we will describe an example of a policy field, to elucidate the options for policy measures and the problems for implementing them.

9.7 Waste: an example of policy options

The increasing volume of solid waste is a matter of great environmental concern. For over a decade the official policy objective in most industrialized countries is that the amount of waste should be reduced and therefore waste management should go up the 'waste management hierarchy' (Wilson, 1996).

9.7.1 The official objectives

For the purpose of waste reduction in the cycle of substances and products the following hierarchy should be adhered:

(1) *Avoid* waste production. The waste that can't be avoided should be
(2) *Minimized*. The minimal amount of waste should go to
(3) *Recycling*. Any waste that is impossible to recycle we first have to
(4) *Treat*. Only then it is fit for
(5) *Disposal*, which should be done with full control and with mitigation of all environmental impact. Controlled and conditioned incineration is mainly preferred to controlled and conditioned landfilling, mainly because of the hazards to the production of drinking water and in densely populated countries also because of the space required.

Practical policy, however, is hardly in line with the objective described in the hierarchy. In most countries of the European Union, for example, landfilling still dominates, while in some countries incineration is fast increasing in significance. Firstly we will briefly summarize options for implementing the official policy. Then some impediments to implementation of the waste management hierarchy are discussed.

9.7.2 Options for policy implementation

For making the official policy objectives effective, several instruments have to be applied in different stages of the product lifecycle. In figure 9.1 the place where the policy instruments listed below should affect the different substance flows are indicated.

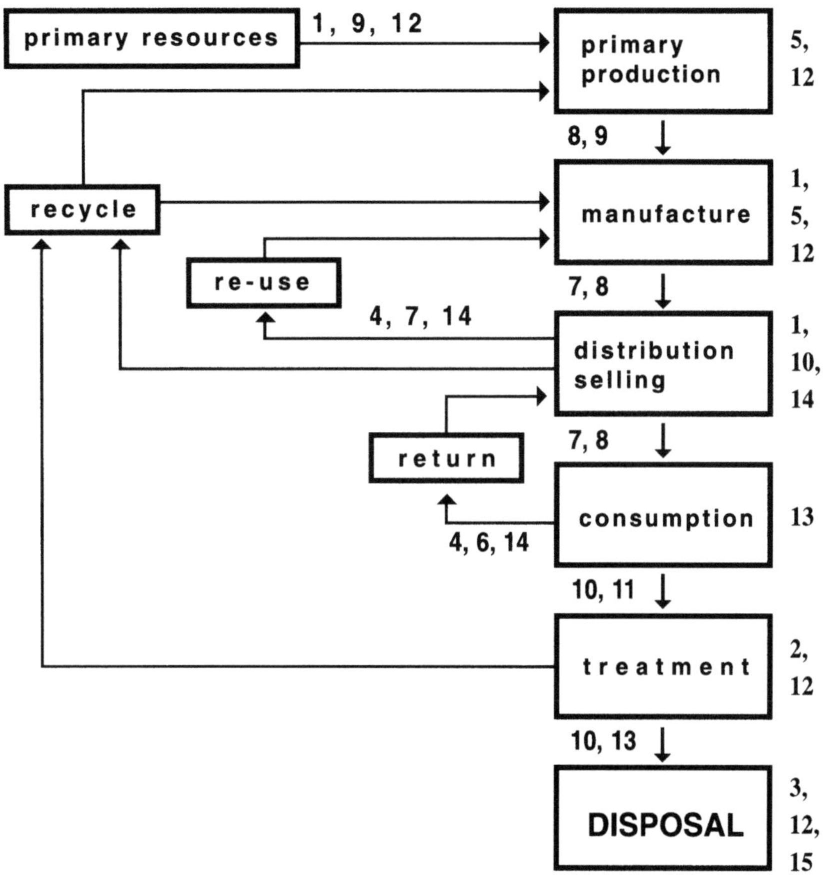

Figure 9.1 Product life cycle and options for policy measures directed at waste reduction.

★ *Coercive instruments* indicated in figure 9.1 involve the following measures aimed at affecting the product lifecycle:

(1) Obligations for internal environmental management and accounting, directed at source waste reduction.

(2) Standards and mandatory monitoring systems of waste treatment.

(3) Standards and norms for and mandatory monitoring of disposal.

★ *Persuasive instruments* affecting the product lifecycle:

(4) Education campaigns directed at co-operation in return and recycling programme's and supply of separated waste.

(5) Facilitating information systems for cleaner production and environmental

management.

(6) Facilitating systems for separate collection for the purpose of re-use, return- and recycling-systems.

★ *Transaction instruments* to be used in the product lifecycle:

(7) Refund systems for product re-use, particularly for packings.

(8) Product responsibility measures for manufacturers.

(9) Taxation (internalizing environmental cost) of the use of primary resources.

(10) Earmarked charges for facilitating return and re-use systems and recycling.

(11) Consumer rates for quantities of waste.

(12) Liability measures for environmental impact.

(13) Taxation (internalizing environmental cost) of waste disposal.

(14) Subsidies for re-use, return and recycling systems.

(15) Abolishing all systems that subsidize waste disposal, including impediments for internalizing external costs, particularly environmental costs.

Depending on the point in the product cycle and on the type of product, different combinations of instruments will be effective. For example, refund systems will only be effective when accessible facilities for refunding are created and educational programmes are supporting public participation. Refund systems alone cannot do the job, and raising awareness with a campaign only will neither be effective. On the other hand, all instruments should also be assessed for their unwanted effects. As described in section 6.7.1 on individual behaviour, for returning particular products or separated waste supply financial incentives may cause lower participation rates or even environmentally damaging ways of litter (Thogersen, 1995).

Nevertheless, all the measures listed above should be considered in order to create a waste management system that really will produce systematic reduction of disposed waste. Although the official primary objective of waste policy is source reduction, only very few measures from the list have been taken in some countries.

9.7.3 Counter movement

In spite of the official objectives, it is waste disposal that generally is still heavily subsidized by society. Waste policy has 'only paid lip service to the hierarchy, acknowledging the supremacy of waste avoidance, minimisation and recycling, while in practice, the vast majority of wastes in all of the EU Member States have either gone to landfill or incineration.' (Wilson, 1996, p.385).

Again, we have to recognize the problem of combining environmental objectives with the more traditional functions of the State. The waste policy instruments indicated in figure 9.1 and listed in the previous section, contain a lot of interventions in economic development. Several restrictions, conditions and new ways of taxation are among the measures to be taken. Since the modern State with its steering and achieving functions is heavily involved in stimulating economic development and is reluctant to take restrictive measures, particularly in the era of growing global competition (chapter 10), many instruments listed are not favoured. Coalitions of actors in the relevant policy domains hold strong policy beliefs directed at avoiding restrictions on economic development, at decreasing costs for business. This explains why technical solutions are so strongly favoured, also in the domain of waste management. There the technical solutions are particularly to be found at the end of the cycle in the stage of treatment and disposal. Besides, all efforts in earlier stages of the product cyles involve a lot more social than technical engineering. It requires more behavioural change, which will always meet resistance, apart from the interest of economic development.

The actor network of the waste sector inherently appears to favour waste disposal and is not suitable for implementing effective Demand Side Management activities. Municipalities invest in disposal plants like incinerators from the perspective of environmental hygiene. Once these facilities are operating their pay back times are highly dependant on full utilization of their capacity. As private enterprises often consider financial risks to be excessive, municipalities or other waste collectors transfer the corporate risks of investments in incinerators to the consumers. Once such investments have been made, any subsequent decrease in waste streams creates a deficit in the exploitation, so there is no more interest in waste minimization (de Jong and Wolsink, 1997).

The fast growing involvement of energy utilities in financial investments for waste facilities is a trend that becomes increasingly significant. Powerful pro-incineration alliances frustrate the attempts to control and decrease the amount of waste in several countries with generating energy as the main argument (Gandy, 1994, p.115). When incinerators are linked to energy generation, the managers of this activity like to define it as a form of processing, although defining it as processing (or recycling) instead of disposal is questionable (Powell, 1993). In Germany, for example, we recognize a strong trend towards the energy utilities getting involved in the disposal and processing side, with enormous financial resources, originally built up for investments in nuclear power. Remarkable for this development in Germany is that the energy utilities are gaining control over the collection side by investing in *Duales System Deutschland* (Dual System Germany). This is a nation wide organization set up by the packaging producers, meant for implementation of producer responsibility for packings (Fishbein, 1994). This organization is showing a strong tendency toward monopoly and has now been taken over largely by energy producers (Benzler *et al.*, 1995). These are eager to gain control over the package waste as a cheap and secure flow of fuel for waste-to-energy facilities.

Against the movement towards investing in disposal instead of waste reduction, a countermovement exists as well. In the United States the problems surrounding the construction of incineration plants have shown that alternative solutions must be taken into consideration by the authorities, if their decision is to be accepted.

It is not sufficient to discuss all the alternatives, which are often aimed at prevention or a reduction in the volume of waste products; the political will to implement them must also be demonstrated. It has been shown that public support for the construction of incineration plants is greater when there is already a waste-reduction programme (Ehrenfeld *et al*, 1989).

There is a great deal of evidence based on research in the United States which points in the direction of low acceptance rates for waste incineration (Rabe, 1991). From the results of survey research on the problems surrounding incineration plants in the United Kingdom, Petts (1992, p.175) concludes that it is only under certain circumstances that the technology of incineration can count on the support of the public. The industry that is involved in waste disposal has an extremely poor image, and its professionalism has also been questioned. In addition, suspicions have been voiced about our present knowledge of the risks represented by the emissions from incineration plants, in particular the long-term health hazards.

9.8 Strengthening openness and democratization

Just as actors are capable of influencing the political process leading up to decisions related to the environment, they can also distance themselves from the implementation of the actual measures. This is one of the causes of ineffective environmental policy. Policy has to be developed, and powerful coalitions are able to influence that process. Then policy has to be implemented, and again many actors have resources to frustrate that process. This all looks rather discouraging, however, as mentioned in the previous section concerning the realization of waste incinerating facilities, policies that include decisions with severe environmental impact also meet resistance within society.

In the case of a decision affecting the environment, the views of a target group may differ considerably from those involved in policy decisions. The problems surrounding nuclear energy plants are a case in point. Attempts by the State to find sites for the storage of radioactive waste quite often encounter a lack of cooperation on the part of the local authorities, and outright opposition from residents (Blowers *et al.*, 1991). Electricity generation by means of nuclear reactors is an unacceptable policy in the eyes of a large proportion of the public in many countries, who see the government not as the guardian of the common good, a mediator, but rather as one of the interested parties.

Friction between the policy of the authorities and the views currently held within a community may also be the result of the separation of the advantages and disadvantages resulting from the measures. It causes the social dilemma connected with environmental problems known as 'LULUS' (locally unwanted land use) and 'NIMBY' (not in my backyard). These are local phenomena which arise when people accept the general usefulness of a planned facility (a railway line, say, or a power plant), but feel that this is outweighed by the disadvantages which must be borne by the local community. These disadvantages happen to be mainly environmental effects. When a permit is issued to drill for oil, that decision is taken for economic reasons, while local opposition arises out of fears about accidents and damage to the surrounding environment (Marks and Von Winterfeldt, 1984).

Counties, towns offered cash to provide nuclear waste sites

The state-run Taiwan Power Co (Taipower) is offering attractive cash rewards for locations within Taiwan which are desperately needed as nuclear waste disposal sites.

Taipower is offering cash rewards up to 3bn New Taiwan dollars (109m US dollars) for any Taiwan county or township which is chosen to allocate a piece of land as the location for Taipower to store low radioactive waste, said Taipower vice-president Kuo Chun-huei. Taipower is also offering 50m New Taiwan dollars (1.8m US dollars) for any local government qualified as candidate, even though it fails in the final selection, Kuo noted. (...). Besides receiving 3bn New Taiwan dollars as feed-back fund, the chosen city will get another 50m New Taiwan dollars as management subsidy, Kuo added. Taipower plans to settle the location scouting in six months, Kuo says.

BBC Summary of World Broadcasts, August 13, 1996.

As noted in section 7.6.2 the public can be motivated to oppose this kind of decisions in four ways. Only one of them, the clear NIMBY type can be overcome by means of economic bargaining. For this type of opposition a market approach is proposed as a strategy, including compensation of local diseconomies (Gregory and Kunreuther, 1990). From a market approach as a starting point the introduction of auctions for finding sites is also suggested (O'Sullivan, 1993). The limited effectiveness of this approach is illustrated by the claim of Inhaber (1997) that auctions are an option to avoid negotiations. However, for the other categories of opponents as distinguished in section 7.6.2 these market strategies will not wipe out the countermovement. Actually, the other three types of opponents outlined in 7.6.2 will feel offended. People who are motivated to oppose for reasons different from the economic rationality of individual benefits minus individual costs, will tend to call it 'bribe'. They are often opposed to the choice of a site, or the design and setup of the facility. In that case the lack of acceptance may be due to the fact that people have had too little influence on the decisions which determine how a certain policy is given concrete form. People feel society is transferring its costs on communities that lack the power to resist, leading to the phenomenon of peripherilization, already mentioned in section 9.3.1. This process may be interpreted as a positive tendency. Gore (1992, p.355) recognises, that in general environmental awareness, also of global environmental problems, begins with the concerns about local environmental impact. He even calls the resistance usually called NIMBY 'a healthy trend' for this reason (Gore, 1992, p.291). The link between environmental impact threatening people's own environment and quality of life, and the environmental values they develop should not be underestimated. Right here all environmental concerns which initially started as not in my backyard feelings are vital. As Hays noted the question: would there be a widespread significant awareness about remote and global environmental issues, when thinking about local threats had not been triggered by initial NIMBY-feelings? (Hays, 1987, p.529).

For this reason the legal rights laid down in public law, created for the improvement of democratic processes, may be the most important instruments for effective environmental policy. Anyhow, as the extensive literature on political decision about risk suggests, improved democracy and openness is the key to sufficient priority for environmental risk (Renn *et al.*, 1995). Legal rights for participation, such as the 'Right-to-Know legislation' are of crucial significance in putting environmental issues on the agenda (Sorokin and Schulkin, 1991). The important lesson is that many rejections of facilities by the public, local pressure groups or even local authorities are not of the mere selfish nature as the phrase not-in-my-backyard suggests. Taking an approach in planning and procedures which tries to counteract pure NIMBY-ism will be counter-productive for all parts of the resistance that have different origins (Wolsink, 1994).

Nevertheless, for at least a part of the population the mere selfish motives may still be dominant, which is certainly creating problems, including for some environmental policy implementation. The type of motives of protest which primarily concern egotistical values may also cause trouble for facilities that serve goals in ecological modernization. In particular when global environmental problems are effectively attacked by some kind of international management, this ecological modernization process requires a protection against the inevitable attack from particular interest groups (Hajer, 1996, p.252). Whether or not the State will be fully equipped for that protection is also a matter of interests and power, in other words, of politics. As we have tried to analyse in this chapter, the State is very much the actor that is essential for improvements in the relation between environment and society, but at present the modern State is not yet fully equipped to deal effectively with all the issues.

Chapter 10
Globalization

10.1 Globalization of environmental problems: the case of air pollution

Environmental problems are more and more defined as global issues nowadays. They are felt and reacted upon everywhere on earth. The number of global environmental actions and global environmental agreements is growing. In chapter 3 we discussed the spatial scale of different environmental problems. However, this is a highly dynamic phenomenon. Nowadays we are witnessing a process of increase of scale of many environmental problems. Problems which were considered as local or regional problems in the past are now dealt with as global problems. Partly this happens because of the transboundary effects of these problems. Environmental effects which are caused in one place are felt and reacted upon in other parts of the world (see section 3.5). Another reason to redefine environmental problems as global issues is the impact of these problems on the complex interdependencies of the planet earth as a whole.

A good example of the globalization of environmental problems is the concern about air pollution in the Netherlands, which was extensively studied by Dinkelman (Dinkelman, 1995). In the seventies air pollution was mainly handled as a regional problem of the Rijnmond area near Rotterdam, damaging the health of the population living in that area. In the eighties air pollution was redefined as a transboundary European problem. At that time, the main concern was acidification, causing among other things damage to forests and lakes. In the nineties global warning became the big issue. CO_2 was added to the list of air polluting gases. Floods, droughts, growing incidence of malaria and other world-wide risks to human well-being were feared as results of higher temperatures on earth.

Three totally different definitions of the problem of air pollution were alternately dominant within a period of 25 years. First the problem was defined in an anthropocentric way: as a health problem. After that it was considered in an eco-centric way: as a problem of dying forests and dying lakes. Then again it was interpreted in an anthropocentric way: as a problem of safety, health and prosperity of future generations of human beings.

This shift in emphasis also implied a change of environmental policy institutions which became involved in the problem of air pollution. Local, regional and national institutions in the Netherlands were involved in air pollution prevention measures in the Rijnmond area. The Economic Commission for Europe (ECE) of the United Nations became the framework in which acidification abatement policies were developed. It initiated a programme for monitoring and measuring transboundary air pollution and prepared the Convention of Long Range Transboundary Air Pollution (LRTAP), which was ratified by countries in eastern and western Europe and northern America (the USA and Canada) in 1979. Within the framework of the LRTAP convention various protocols were signed on different pollutants. In 1985, 18 countries signed the Helsinki protocol and thereby committed themselves to reducing SO_2 emissions by 30% by 1993, compared with the level of emissions in 1980. Under the 1988 Sofia Protocol, the signatories were committed to stabilising NO_x emissions to 1987 levels by the year 1994. A third protocol, signed in Geneva in 1991, aimed to limit emissions of volatile organic compounds (VOCs) in 1999 to a level of 30% below that applying in 1988 (Hordijk, 1991; OECD, 1993a pp.179-181). All western countries and most of the eastern European countries had reached the 30% SO_2 reduction target in 1993. Many western countries achieved a much larger reduction in SO_2 emissions (Tellegen, 1996, p.82).

The increase of scale and globalization of environmental problems has not only been limited to air pollution. Water pollution, hazardous waste and nuclear and other industrial accidents are examples of other issues that have become matters of global environmental concern. Today three world-wide environmental issues rank high on the global policy agenda: depletion of the ozone layer, climate change and reduction of biodiversity. We will shortly describe their physical aspects and discuss their social implications in section 10.3. We precede our presentation of these global problems with some remarks about the physical properties of the planet earth.

10.2 The planet earth: what kind of a system?

The growing human impact on the planet earth has been well documented by scientists since a long time. In 1864, the American geographer George P.Marsh published his book 'Man and nature; or, Physical Geography as modified by human action' (Marsh, 1864). Marsh described the impact of human activities all over the world and concluded that there are no longer places left on earth which are free of human influence. The message of his book was summarized on the title page by a quotation from the protestant minister Horace Bushnell: 'Not all the winds and storms, and earthquakes, and seas, and seasons of the world, have done so much to revolutionize the earth as Man, the source of an endless life, has done since the day he came forth upon it, and received domination over it' (Marsh, 1864, p.1). More than half a century ago the Russian scientist Vernadsky argued that human activities are changing the 'biosphere' into a (man-made) 'noosphere'. He called the noosphere a new geological phenomenon on our planet in which man becomes the largest geological power (Vernadsky, 1945 and 1991, p.241). In the twentieth century human impact on life on earth has been enormously enlarged. A growing share of the earth's biomass is directly used by human beings (Vitousek *et al.*, 1986).

We are well-informed about all sorts of human influences on living and non-living elements of the earth. However, we know far less about the long-lasting effects of these anthropogenic changes. The reason for this is a lack of understanding of the kind of relationships that exist between different living and non-living elements of the planet earth as a whole. Many years ago Boulding presented 'a hierarchy of complexity of general systems', starting with frameworks as the most simple type of systems and followed, in sequence of growing complexity, by clockworks and the thermo-stat. After that followed 'open systems' or 'self-maintaining structures', of which flames and rivers were mentioned as simple forms. Boulding continued the presentation of his hierarchy of systems remarking that 'as we pass up the scale of complexity of organization towards living systems, however, the property of self-maintenance of structure in the midst of a throughput of material becomes of dominant importance' (Boulding, 1968, p.7). We know of all sorts of self-maintaining mechanisms in human and animal bodies. The concept of homoeostasis refers to the processes by which bodily temperature, concentrations of substances and other equilibria are maintained under changing external circumstances. Boulding designed his hierarchy of system models in order to 'prevent us from accepting as final a level of theoretical analysis which is below the level of the empirical world which we are investigating' (Boulding, 1968, p.9).

In 1969 Lovelock put forward his Gaia hypothesis which implied that we have to interpret the functioning of the planet earth with a system model of higher complexity than we are used to do. The name Gaia or 'Mother earth', indicating that earth is a living entity, was being already used by the Greeks 2000 years ago. Lovelock defined Gaia as 'a complex entity involving the Earth's biosphere, atmosphere, oceans, and soils; the totality constituting a feedback or cybernetic system which seeks an optimal physical and chemical environment for life on this planet' (Lovelock, 1988, p.11).

The Gaia hypothesis has been criticized because of internal contradictions and lack of empirical support (Schneider, 1990). However, global negative (as well as positive !) feedback mechanisms do exist. Neither can it be denied that, owing to a basic lack of understanding of the functioning of the planet earth, global risks like those discussed in this chapter are surrounded by scientific uncertainties.

10.3 Three global risks.

At the end of the twentieth century three phenomena are considered as major global environmental problems: depletion of the ozone layer, climate change and reduction of biodiversity. Of these three problems the first two are defined as disturbances of the planet earth as a whole. The third one is more an 'umbrella' problem embracing problems which have been defined and reacted upon at local and regional levels for a long time. What are the main characteristics of these global problems and how are they dealt with?

10.4 Ozone depletion

10.4.1 Causes and effects

The ozone layer, located in the stratosphere, protects life on earth against harmful solar radiation. Since about 1970 the world has been alarmed about the global risk of depletion of the ozone layer. Reduction of the ozone layer is expected to cause cancer of the skin and to have damaging effects on agricultural productivity and the marine food chain (Benedick and Pronovo, 1992, p.137).

At first it was argued that supersonic transport planes like the Concorde, flying in the lower part of the stratosphere, could damage the ozone layer by their emissions of NO_x (Crutzen, 1971). Then the American researchers Molina and Rowland discovered another threat to the ozone layer. Chlorofluorcarbons (CFCs), used in among other things refrigerators and sprays, ultimately reach the stratosphere. There they are broken down by ultraviolet radiation, whereby atomic chlorine is released. Chlorine as a catalytic agent breaks ozone (O_3) down into oxygen (O_2). The researchers concluded that, if trends in the emissions of chlorofluorcarbons do not change, the ozone layer will be depleted in the foreseeable future (Molina and Rowland, 1974). Increasing loss of ozone over the Antarctic and Arctic region has been confirmed since then time and again, though it was not always clear to what degree this was caused by human activities or by natural causes like the eruption of volcanoes.

10.4.2 Policy measures

After the publication of the article of Molina and Rowland, hearings were held in the US congress in 1975. Three years later the use of sprays containing CFCs was prohibited in the US.

The first result of global policies to protect the ozone layer was the 'Vienna Convention for the protection of the Ozone layer' of 1985. This convention was intended to stimulate international policy measures and research efforts. In 1987 it was followed by the 'Montreal protocol on substances that deplete the ozone layer' which was signed at that time by 24 countries. From now on participating countries agreed to reduce the use of ozone depleting gases. Further decision-making took place in annual meetings of the parties to the Montreal Protocol and biennial conferences of parties to the Vienna Convention (Benedick and Pronovo, 1992).

10.4.3 Success and failure

Global policy measures on prevention of depletion of the ozone layer are considered as one of the greatest, if not the greatest, successes of global environmental diplomacy (Benedick, 1991; Gijswijt, 1996). The use of most of the ozone damaging gases like CFCs and halones has gone down in many countries. Part of the world-wide success of these efforts can be contributed to the special arrangements for developing countries. They were allowed periods of delay before the total abandonment of ozone depleting gases and received both technical and financial support from developed countries in order to be able to implement policy measures agreed upon.
International political decisions have contributed to the stabilisation and reduction of emissions of ozone depleting gases. Given the condition that current trends continue, in the middle of the twenty first century the ozone layer is expected to have the same thickness as it had in the middle of the twentieth century, before human activities started to have damaging effects.

10.5 Climate change

10.5.1 Causes and effects

In countries with temperate and cold climates fruit and vegetables are cultivated in greenhouses. Their walls and roofs of glass transmit solar radiation but prevent the heated air escaping. In the atmosphere water vapour (H_2O), carbon dioxide (CO_2), nitrous oxide (N_20), ozone (O_3) and other so-called natural greenhouse gases function like glass in greenhouses. These gases permit solar radiation to reach the earth's surface but hamper the ascent of heat to higher layers of the atmosphere (Van Beukering and Vellinga, 1996, p.188). Without this so-called 'natural greenhouse effect' temperature on earth would be about 33° Celsius below the actual level.
As early as 1896 the Swedish scientist Arrhenius suggested that the large-scale burning of fossil fuels could lead to a rise of carbon dioxide in the atmosphere and thereby raise the surface temperature of the earth (Brenton, 1994, p.163). Later it was discovered that other gases like methane (CH_4), nitrous oxide (N_20) and (chloro)fluorocarbons (CFCs) which are emitted as a result of human activities also strengthen the greenhouse effect of the atmosphere. This

phenomenon has become known as the (enhanced) greenhouse effect. The sources of greenhouse gases are very diverse (see table 10.1).

Table 10.1 Relative direct sources of global greenhouse gases (percentages of total anthropogenic emissions in 1990)

Sources	CO_2	CH_4	N_2O	CFCs
Energy	80	26	9	-
Deforestation/land clearing	18	-	17	-
Other industry	3	-	15	100
Fertilised soils	-	-	48	-
Enteric fermentation	-	24	-	-
Rice cultivation	-	17	-	-
Landfills	-	11	-	-
Biomass burning	-	8	11	-
Animal waste	-	7	-	-
Domestic sewage and	-	7	-	-

Source: IPPC (1992); Stanners and Bourdeau (1995, p.515.)

There can be no doubt about the rising concentration in the atmosphere of the above mentioned greenhouse gases as a result of human activities during the last two centuries. However, there is no one-to-one relationship between these higher concentrations and rising temperatures on earth. Burning of fossil fuels does not only result in the emission of carbon dioxide but also contributes to the formation of aerosols in the atmosphere, which reflect solar radiation and influence the formation of clouds and thereby reduce the global warming effect of the emission of greenhouse gases. Apart from that, climatic change is not only dependent on atmospheric circumstances but also on biological factors, like the state of tropical forests and the presence of phyloplankton in oceans and their capacity to absorb carbon dioxide. The greenhouse effect might be enhanced by positive feedbacks and mitigated by negative feedbacks. An example of the former would be a rise in temperature, caused by the greenhouse effect, leading to the release of large quantities of methane as a result of the thawing of tundra soils and in that way strengthening the greenhouse effect. An example of a negative feedback mechanism is plant growth caused by rising temperatures and resulting in higher absorption of greenhouse gases (Van Beukering and Vellinga, 1996, p.191).

Notwithstanding the above mentioned uncertainties there is a growing consensus among scientific researchers, cooperating in the intergovernmental panel on climate change (IPPC), that larger concentrations of greenhouse gases are already leading to a rise of temperatures on earth now. Rising sea levels and floods, local water shortages and desertification, extreme weather events (in particular storms) and growing incidence of the tropical illness malaria are among the expected effects of climatic change.

10.5.2 Policy measures

Whereas in the case of the shrinking of the ozone layer it took less than ten years before the warnings of scientists were followed by policy measures, in the case of the enhanced greenhouse effect the period between the first scientific message and the first policy measures lasted nearly a century.

In 1985 the United Nations Environmental Programme (UNEP) discussed the role of greenhouse gases in causing climate change with the World Meteorological Organization and the International Council of Scientific Unions (Van Beukering and Vellinga, 1996, p.198). The publication of the Brundtland report in 1987 placed the issue of climate change high on the global environmental policy agenda. In the US the hot and dry summer of 1988, with its disastrous effects on agriculture, strengthened the belief in the existence of an anthropogenically enhanced greenhouse effect. Also in 1988 300 experts present at the Toronto conference on the changing atmosphere recommended that countries should reduce CO_2 emissions by approximately 20% of 1988 levels by the year 2005. In the same year the already mentioned Intergovernmental Panel on Climate Change (IPPC) was established by WMO and UNEP. IPPC became the world-wide platform for exchange of knowledge on climatic change and design of policy measures to reduce climatic effects of human activities. Climate change was one of the main issues of the United Nations Conference on Environment and Development in Rio de Janeiro in 1992. During that conference the United Nations Framework Convention on Climate Change (FCCC) was signed by 155 states and the European Union.

10.5.3 Success and failure

The UN convention does not contain strict targets and timetables. The ultimate objective of the Convention is to 'achieve stabilisation of greenhouse gas concentrations in the atmosphere at a level that would prevent dangerous anthropogenic interference with the climate system within a timescale sufficient to allow ecosystems to adapt naturally to climate change, to ensure that food production is not threatened and to enable economic development to proceed in a sustainable way' (Stanners and Bourdeau, 1995, p.520).

This convention does not even require a stabilisation of emissions of greenhouse gases in the developed countries which emit the most. Unlike the United States, which opposed stronger policy measures at that time, the European Union decided to strive for stabilisation of emissions of the most important greenhouse gas CO_2 at the 1990 level in the year 2000. Some member states of the EU were even more ambitious and declared the aim of a reduction of total carbon dioxide emissions in their country. Probably the policy target of stabilisation of CO_2 emissions will not be reached by the European Union and other developed countries. In Eastern Europe the economic decline during the transformation from a planned to a market economy has led to a temporary decline of CO_2 emissions. However, it is highly predictable that as soon as this period of decline is followed by a new period of economic growth CO_2 emissions will rise again. In other transition countries like China and the new industrial states in south-east Asia, CO_2 emissions have been rising rapidly in recent years. After the United States, China became the country with the second largest absolute CO_2 emission in the world.

Emissions of the most important greenhouse gas CO_2 are caused by the burning of fossil fuels for energy generation and thereby related to different core activities in developed and developing countries. Measures to reduce carbon dioxide emissions therefore meet with strong opposition from vested interests. Notwithstanding the fact that there are many possibilities for cost-effective energy conservation measures, states often argue that measures within one country damage national economic interests and hardly contribute to the solution of the problem: a clear example of a social dilemma (see chapter 5).

Apart from that, the negative effects of anthropogenic climate change are not all felt to the same degree in different parts of the world. Some cold areas like Siberia may even profit from climate change and expect larger agricultural yield. Small island states, like the Maldive Islands in the Indian ocean, which run the greatest risk of floods caused by a rising sea level, are now united in the Alliance Of Small Island States, AOSIS. They are demanding strong, world-wide policy measures in order to reduce the enhanced greenhouse effect. However, without the support of other states they are unable to defend their vital interests in the prevention of a catastrophic rise of sea level. Because of these circumstances and other factors mentioned above, the inability to prevent climate change might become one of the greatest failures of global environmental policy in the near future.

10.6 Reduction of biodiversity

10.6.1 Causes and effects

Life on earth manifests itself in a great variety of ecosystems, species and genes. This so-called biodiversity is a dynamic phenomenon, which is strongly influenced by both natural circumstances and human activities.

Numbers of species living on earth as well as data on anthropogenic extinction of species can only be presented as rough estimates. Estimates of the total number of species living on earth today run from 5 to 15 million. The rate of extinction of species *per annum* is estimated to be 10.000 to 20.000 (Barnes, 1996, p.221). In the past, natural events have led to a, sometimes dramatic, reduction of biodiversity. Human activities may contribute to growth as well as to reduction of biodiversity. Extensive forms of agriculture have contributed to growing biodiversity. Today intensive agriculture is one of the main causes of reduction of biodiversity.

'Reduction of biodiversity' functions as an umbrella concept embracing differing, widely shared, environmental concerns. In general, people are not alarmed by the disappearance of 'ecosystems' and 'species' but they do have strong feelings about the disappearance of tropical forests and of particular species of animals and even plants, if not on a world-wide-scale, then at least in their own country or region. 'Loss of genetic diversity' will be regarded as a highly abstract notion by most people. However, there are great commercial interests of biotechnological enterprises involved in the protection of this form of biodiversity.

10.6.2 Policy measures

In the past, 'loss of biodiversity' was not a main issue of national or international environmental activism and policy making. But in 1992 a convention to protect biodiversity was one of the main outcomes of the UNCED in Rio de Janeiro. More concretely formulated as the care of single species of animals and plant protection of biodiversity has a much longer international history. International protection of endangered animal species began in the nineteenth century.

History of protection of endangered animals

It was in 1872 that the Swiss first proposed the establishment of an international commission to protect migrating birds. Probably the first ever international agreement was the 1900 Convention on the Preservation of Animals, Birds and Fish in Africa, signed in London by the European colonial powers with the (strikingly farsighted) intention of preserving game in east Africa by limiting ivory exports from the region. The late nineteenth and early twentieth century also saw an international convention to protect fur seals, an agreement among littoral states on the management of the Rhine and a US-Canadian agreement on the protection of migrating birds.

Brenton, 1994, pp.15-16.

A hundred years after the first initiatives on international protection of endangered animal species, new international agreements were made. Of great importance were:

- The 1971 Ramsar Convention on Wetlands of International Importance especially as Waterfowl Habitat.
- The 1973 Washington Convention on International Trade in Endangered Species of Wild Fauna and Flora (CITES).

The Ramsar convention and CITES are illustrative of two different ways of protecting endangered species of animals. CITES illustrates a direct approach, aiming at the prevention of hunting or gathering. The Washington convention is an example of the indirect approach, protecting the habitats of endangered animal species. These two strategies are a reaction to the direct and indirect exhaustion of renewable resources which we distinguished in section 1.3.1.

Wetlands are an example of a type of habitat that is extremely important for the survival of endangered species. Tropical forests are another example. Huge economic interests are involved in the exploitation as well as the conservation of tropical rainforests.

Tropical forests are of global environmental interest because of their rich biodiversity and their crucial role in regulating CO_2 concentrations in the atmosphere (Kolk, 1996). But unlike the ozone layer, the atmosphere and the oceans, they are located within the boundaries of national states. 'Forests are thus both a 'global commons' providing a collective good from which all benefit and the 'property' of an individual state' (Hurrell, 1992, p.402).

Destruction of tropical forests does not only lead to reduction of biodiversity but also contributes to the enhanced greenhouse effect as a result of the release of carbon dioxide. The causes of loss of tropical forests are manifold and vary widely between different countries like Brazil and Malaysia. Both subsistence farming and harvesting of wood for fuel by the local population and wood production and agriculture for export contribute to it.

10.6.3 Success and failure

Previously (see section 10.3) we called biodiversity an umbrella concept embracing many different forms of natural degradation which have been known for longer and are easier to react upon than the highly abstract phenomenon of 'reduction of biodiversity'.

For a long time, there has been great public support for protection of endangered wildlife. Organizations operating world-wide like Greenpeace and the World Wildlife Fund were highly successful in their struggles for the protection of what has been called 'charismatic megafauna': a small number of animal species like seals, whales and elephants, which arouse strong emotions, in particular among the television watching audience in developed countries (Hannigan, 1995, p.154). A much wider range of endangered species is covered by the previously mentioned CITES treaty, which is considered as a rather successful global environmental treaty.

Its successes are based on the common interests of developing and developed countries. The former are interested in the protection of their indigenous species (and major tourist attraction) against poaching and illegal trading, the latter are pressed by their electorates to contribute to wildlife conservation (Brenton, 1994, pp.101/102).

Far more controversial than the protection of existing species is the question of ownership of new species which are developed by biotechnological enterprises. Controversies about this issue were the cause of the USA withholding its support to the biodiversity convention in Rio de Janeiro in 1992.

Commercialisation of biodiversity

One very clear example of the commercialisation of nature is the situation with seeds. Throughout the history of humankind, seeds have always been accessible to farmers. Until quite recently, small business or government agencies concerned with improved productivity in domestic agriculture were in charge of the seed supply farmers. However, since the development of hybrid seeds (which are developed in laboratories and do not allow for replanting of harvested seeds), the way has been clear for big business to control the market. A private company can now own a seed variety and prohibit others from copying it. In order for major capital interests to control the seed market, mechanisms were need to control, dominate, appropriate, and own this natural good, the seed. This is now being made easier by the patenting mechanisms. (...)
Just as with seeds, the generic diversity of the Southern countries (particularly those situated in the Amazon Region) has become a strategic resource, since it has enormous potential for meeting demands of the industry in the First World. It is no coincidence that big capital interests are now manifesting their 'concern' over the biodiversity that still exists on the planet, even as they step up the struggle to take possession of these resources.

Couto Soares (ed.), 1992, p.18.

In order to protect tropical forests, being the habitat of an extremely rich variety of species, cooperation between developing countries which produce and developed countries which consume tropical timber is necessary. They do work together in the International Tropical Timber Organization which is based on the International Tropical Timber Agreement (ITTA) of 1983. ITTA does not contain clearly circumscribed obligations to protect tropical forests and to compensate financial losses as a consequence of reduced export of tropical hardwood. Until now this agreement and other forms of international cooperation have not been effective in stopping the destruction of tropical forests.
Protection of biodiversity is put into practice by the protection of limited numbers of ecosystems or species. Only particular areas and particular species are protected, instead of natural variety as such.

10.7 Global environmental actors

10.7.1 The construction of environmental problems

In chapter 1 of this book we stressed the fact that man-made environmental

changes only become environmental problems when they are considered as problematic by human beings. In other chapters we discussed the way in which human induced environmental changes are reacted upon by individuals (chapter 7) and political systems like states (chapter 9). Without denying the damage done to nature and human beings by anthropogenic environmental change, one may argue that environmental problems are 'constructed' by human beings. In his constructionist study of environmental problems Hannigan distinguishes different tasks in the construction of environmental problems which are performed by different groups. Scientists play a major role in the discovery and definition of problems, the mass media arouse attention among the public at large and politicians get the problems listed on the political agenda (Hannigan, p.42).
This characterisation of the construction of environmental problems does not fully apply to small-scale environmental problems. The involvement of scientists and journalists is not always required to 'construct' local environmental problems like the nuisance from noise, stink and dangerous traffic and the regret about the felling of trees in the local neighbourhood. However, in the discovery and the arousal of concern about human-induced global changes both science and mass media play a crucial role. The IPCC is a clear example of the crucial role of scientists in defining climate change as a global environmental problem. (Smith, 1996). Mass media play a crucial role as well in mobilizing support for global environmental problems, protection of 'spectacular' endangered species and in the visualisation of 'the hole in the ozone layer'. On the other hand, scientific controversies and declining interest of the mass media can also contribute to what could be called the 'deconstruction' of environmental problems. The functioning of global networks of scientists and the global spread of information by mass media play a crucial role in the process from signaling to the solving of global environmental problems. However, scientists and journalists are not the only actors who play a role in this process. There are other important actors as well such as public authorities, private firms and non-governmental organizations.

10.7.2 Public authorities

Different public bodies, varying from world-wide organizations to municipalities, are or may become engaged in global environmental issues. Important world-wide operating organizations include the United Nations and its specialized departments like UNEP (United Nations Environmental Programme) and UNDP (United Nations Development Programme) and financial organizations like the World Bank and the International Monetary Fund. Other organizations with only a limited number of member-states, like the OECD (Organization for Economic Cooperation and Development), consisting mainly of western industrialized nations, and the European Union also play a role in global environmental policies. Even local authorities like municipalities may become involved in global environmental issues. An example of local involvement is the banning of the use of tropical hardwood for building by municipal authorities in wood importing countries.

10.7.3 Private firms

Private firms, varying from huge multinationals to small firms of consultant, play an important role in the handling of global environmental problems.

Often enterprises will resist global policy measures because they force them to reduce their production or to invest in cleaner production processes. Some enterprises, however, may profit from global environmental policy measures. Policies to counteract the enhanced greenhouse effect by reducing CO_2 emissions will offer new opportunities to enterprises which specialize in renewable energy technologies. These enterprises, working together in the 'Business council for a Sustainable Energy Future', support global policy measures to reduce the emissions which are caused by the use of fossil fuels. Another branch of private business which strongly argues for policies preventing climate change is the insurance sector. Insurance companies fear large financial losses as a consequence of extreme weather conditions caused by climate change.

10.7.4 Non-governmental (environmental) organizations (NGO's)

An important role in global environmental issues is played by world-wide operating environmental organizations. Three of the most important of these organizations are the IUCN, Greenpeace and Friends of the Earth.

- The International Union for the Conservation of Nature (IUCN) is a world-wide organization of what might be called a 'hybrid' nature. Governments, as well as publicly sponsored institutions like universities and non-governmental organizations, like Greenpeace, are members of IUCN. Nature conservation is the main field of action of IUCN.
- Greenpeace became famous for its spectacular actions at sea against the killing of whales. Today it fights many other damaging activities at sea (like the extraction of oil and gas and the dumping of radio-active waste) and on land (like testing nuclear bombs and the destruction of forests). The organization not only tries to stop polluting industries, but also to stimulate clean industries. For example, Greenpeace became a promoter of the CFC-free refrigerator of the Scharfenstein company in Chemnitz in the former GDR. Greenpeace combines spectacular actions against concrete environmentally damaging activities with less spectacular research and lobbying activities.
- Friends of the Earth was founded in the USA in 1969 and is now an international environmental organization with national branches in many countries. Whereas Greenpeace is a centralized organization with local offices in different countries, Friends of the Earth is a decentralized organization in which national member-organizations participate with their own specific goals and strategies.

Many other types of organizations, varying from public utilities to universities,

incidentally or permanently become involved in global environmental issues. A wide range of interactions has developed between players of the same type (for example between world-wide operating NGO's and local groups struggling for the protection of tropical forests) and players of different categories (for example NGO's and mass media).
More and more environmental organizations are engaged in the formulation and implementation of public international environmental agreements.

10.8 From signalling to problem-solving

It would be far beyond the scope of this book to describe the development of the manifold networks between the different players and their impact on global environmental issues, even if we would limit ourselves to the three global environmental problems mentioned in section 10.3. Instead of that we will formulate ten questions which are relevant for the understanding of the fate of global environmental issues between signalling and problem-solving. The questions will be answered briefly and provisionally for the three global issues which were discussed in section 10.4.

a. Has the existence of the problem been proven scientifically?

In answering this question we are not primarily interested in the contents of scientific journals but in the presentation of scientific findings to the public involved. Nowadays it seems that ozone depletion is less surrounded with scientific controversies than climate change. In the latter case the public is much more informed of uncertainties and controversies between scientists regarding relationships of cause and effect. In the case of the 'umbrella' phenomenon of reduction of biodiversity the public is informed with a mixture of clear numbers of single species and vague estimates of overall biodiversity

b. Is the problem presented by the mass media in a way that arouses public attention?

The mass media play a crucial role in world-wide actions to protect endangered animal species like whales and elephants and ecosystems like tropical forests. In the latter case a tragic event like the killing of Chico Mendes (see chapter 8) stimulated media coverage of the struggle to protect nature and people. The visualisation of 'the hole in the ozone layer' by mass media also supported world-wide environmental policy measures in this field. Climate change is a much more difficult phenomenon to present in such a way that it will arouse the attention of a large public.

c. Is it clear that the problem is caused mainly by human activities?

Although it is known that eruptions of volcanoes contribute to the depletion of the ozone layer there is no longer much doubt about the contribution of human activities to this clearly observed phenomenon. In the case of the enhanced greenhouse effect there is no doubt about the contribution of human activities to growing concentrations of greenhouse gases but it remains doubtful whether, and to what degree, these growing concentrations actually contribute to world-wide climate change. There is little doubt about the destructive effect of human activities on biodiversity in general, but it is unclear which activities contribute to what degree and to which form of reduction of biodiversity.

d. Does the phenomenon have mainly negative effects?

In the case of depletion of the ozone layer, the answer is affirmative. Although even in this case there will be some who profit from the need for protection against ultraviolet radiation. In the case of the enhanced greenhouse effect and the reduction of biodiversity there are winners and losers. A warmer climate will lead to higher agricultural yields in the colder parts of the earth. Both farmers and fishermen may profit from 'reduction of biodiversity' in the form of the disappearance of certain types of birds.

e. Are the negative effects uniformly felt all over the world?

The answer is no for all three phenomena. Ozone depletion in particular is a problem of the arctic and antarctic regions and not of the tropical parts of the world. But it implies health risks in the influential rich countries of the world. Global warming will have widely differing climatic effects in various parts of the world. Biodiversity may be growing, instead of shrinking, in many places. For example, in some cities species of birds and plants appear which were never seen there before.

f. Are there negative health effects?.

The answer is yes in the case of ozone depletion, but less evident in the other two cases. Growing incidence of malaria is mentioned as a possible consequence of climate change (Zwick, 1997) but plays a minor role in the presentation of the problem to the larger public. Loss of biodiversity may limit the development of new medicine, but is not in the first place defined as a health problem.

g. Is the problem caused by a small number of players?

Compared to the other two issues, ozone depletion is caused by a relatively small number of human activities. A limited number of enterprises, like producers of insolation materials, refrigerators and sprays, had vested interests in the use of CFC's. Climate change is related among other things to the enormous variety of activities in which fossil fuels are used and to methane from agricultural production. Oil producing countries, industrial enterprises, airlines and the car industry have powerful lobbies to defend their interests and to prevent measures to reduce CO_2 emissions. The variety of human activities which contribute, directly or indirectly, to the disappearance of species is even greater than in the case of climate change. The power of players who want to block biodiversity prevention measures will vary between different protected species and ecosystems.

h. Is it possible to avoid or to change activities which contribute to the problem at little cost and effort?

Once more the answer is much more positive in the case of ozone depletion than in the case of climate change and biodiversity loss. It was comparatively easy and cheap to develop alternatives for CFC's using in sprays and refrigerators. Measures to solve the other two problems require far more cost and effort, although, in order to prevent global warming, there are still many unused methods to reduce energy consumption in a profitable way. Both in the case of climate change and biodiversity loss this question has to be specified for various activities contributing to these two global problems.

i. Are those who (will) suffer most from the problem powerful?

Wealthy captains of industry and powerful political leaders can become victims of the increased risk of cancer of the skin caused by ozone depletion. It has been suggested that president Reagan's support for strong global measures to prevent ozone depletion was stimulated by the fact that he himself had undergone removals of two skin cancers in 1985 and 1987 (Benedick, 1991a, p.67). Both in the case of climate change and biodiversity loss personal risks for powerful people are much more limited. In the case of climate change powerful heads of insurance companies and banks may fear substantial losses as a consequence of less stable weather conditions. Biodiversity loss may be feared by biotechnological enterprises and the pharmaceutical industry. In general the comparatively powerless people in developing countries are much more directly dependent on climatic and biological conditions in their living area than the more powerful people in developed countries. Destruction of tropical forests deprives native people of food and shelter. Floods caused by global warming will take the heaviest toll of human life in powerless developing countries like Bangladesh.

j. Are there vested interests in signalling and solving the problem

Non-governmental environmental organizations, governmental environmental agencies and research institutions have vested interests in the existence of environmental problems. For example, the leading role of the USA in world-wide ozone depletion prevention policies was encouraged by research interests of the National Aeronautics and Space Administration (NASA). Other organizations can be added to this list. For each single environmental problem the question may be asked who is interested in the 'construction', the continuation and the solving of the problem.

The interest of particular organizations in specific issues may change in the course of time. In section 10.2. we referred to Dinkelman's study of air pollution policies in the Netherlands. In the eighties, climate change was neglected by the Dutch ministry of the environment because the problem could not be solved by means of the prevailing 'end-of-pipe' solutions. Ten years later, this disadvantage became an advantage. Abatement of the greenhouse effect became the yardstick for stimulating energy conservation and other preventive measures, like cleaner production, which were required to attain new, more ambitious environmental policy goals which could not be reached by means of 'end-of-pipe' measures.

10.9 Environmental protection and other priorities

10.9.1 Global conflicts of interests

Environmental problems are felt all over the world. According to the results of world-wide research, they are 'salient and important issues in both wealthy and poor nations, and residents of poor nations express as much concern about environmental quality as do these living in wealthy nations' (Dunlap *et al.*, 1993, p 36). Environmental problems rank high on global policy agendas. However, there are other global priorities which are difficult to combine with environmental goals. We will shortly discuss three global policy goals which are often in conflict with environmental goals.

10.9.2 Development and growth

The global environmental crisis was 'constructed' in the first world. At the first UN environmental conference in Stockholm in 1972 both second and third world countries criticized the first world countries for misinterpreting a problem of the rich, capitalist countries as a problem that also involved developing and communist countries. In particular the message of the famous report to the Club of Rome, that was also published in 1972, that there are 'limits to growth' was vigorously rejected by representatives of second and third world countries.
The controversy between the capitalist and communist world has largely become a historic phenomenon, but the conflicts of interest between 'environment' and 'development' are still alive. Both the publication of the report 'Our common future' of the Brundtland commission in 1987 and the second world-wide environmental conference of the UN in Rio de Janeiro in 1992, were intended to bridge the gap between 'environment' and 'development'. The name of the conference was 'United Nations Conference on Environment and Development'. It has often been argued on these and other occasions that development based on economic growth does not hinder environmental protection but that it is even necessary to finance necessary technological improvements.
If this were true the highest 'developed' countries would be the most successful protectors of nature and environment. However, in general they emit more pollutants and use more scarce resources *per capita* than the less developed countries. A good example of this is CO_2 emissions. In general, rich countries use energy more efficiently and therefore emit less CO_2 per unity of the gross national product than poorer countries. However, as a consequence of their high consumption level, the levels of CO_2 emissions *per capita* are higher in the rich than in the poor countries.

10.9.3 Free trade

Free trade is a major global policy goal. The World Trade Organization (WTO) was founded to strengthen free trade in goods all over the world. Nature and environmental quality may profit from free trade. This will be the case when free trade stimulates that production is concentrated in places where it has the least environmental impact and when these environmental benefits are larger than the negative environmental impact of world-wide transport to consumers. The ideal situation is often far away from reality.
In general, agricultural and industrial production will be cheaper in places with weak than in places with firm standards and strong environmental control. For that reason, free trade will stimulate production in places where it has the largest negative environmental side-effects. When this happens, regions with strict environmental control are tempted to soften their environmental regulations in order to protect their economic activities and to stimulate investment of existing and new enterprises (Daly, 1996). In international agreements this dilemma is not always solved. Some environmental agreements contain sanctions against environmentally damaging products which are in conflict with the principles of free trade (Reijnders, 1996[a], p.3).

UN conference backs 'trade measures' to aid environment

Trade sanctions should be allowed in some cases to enforce world environmental policy, ministers and experts from 43 nations agreed on Friday in a declaration issued at the end of a two-day conference in Geneva. The sanctions issue was one of the most controversial at the international Trade, Environment and Sustained Development meeting organized by the U.N. Conference on Trade and Development (UNCTAD) and the U.N. Environment Programme (UNEP). Ministers agreed that 'trade measures' were legitimate in some cases to back up international agreements but did not specify if these ought to include boycotts of products or other cut-offs in trade

Deutsche Presse-Agentur, October 1, 1996.

10.9.4 Autonomy

Many of today's developing countries were colonies until the middle of the twentieth century. They have gone through often painful processes of colonisation and de-colonisation. Oppression by western governments, churches and enterprises was one aspect of western colonisation. Glorification of 'natural' living conditions of native people, unspoiled by western civilisation, another one (Grove, 1995). Nowadays, western interference with environmental issues in third world countries is experienced as neo-colonialism in these countries. 'Ecototalitarism' and 'ecopopulism' are terms which are used to characterize new forms of control from above and glorification from beneath by the west (Dietz, 1996). Western initiatives to protect nature in developing countries are often in conflict with the interests of people living in these countries. A former member of the European Parliament analysed measures of the European Parliament and other international authorities to protect seals in Greenland, kangaroos in Australia, bears in the Beringsea and elephants in Africa (Muntingh, 1995). Prohibition of hunting of bears, of trade in ivory and of import of seal fur and of most species of kangaroos were the main policy measures to protect the endangered species. However, in some cases they also caused mass unemployment and alcoholism, drug addiction and suicides among the local hunting populations. In Africa, the prohibition of ivory trade has led to an excessive growth in the number of elephants in some countries. They are destroying the fields and houses of the local population and even their own habitat. Moreover the income of ivory trade is no longer available for investment in game parks and nature reserves. World-wide interference with local affairs in order to protect nature and environment runs the risks of damaging other local interests and, because of unanticipated side-effects, even nature and the environment.

10.9.5 Who against whom?

The last dilemma, discussed above, suggests that we have to do with conflicts of interests between first world dominated international authorities and national and local authorities in developing countries or poor regions. However, reality is far more complex. Many players are involved in world-wide environmental relationships and in world-wide dilemmas between environmental protection and other political priorities. Take for example the transport of hazardous waste. Hazardous waste is not less hazardous in countries where it is imported than in countries from which it has been exported. On the contrary, in particular chemical waste has often been exported from countries with comparatively strict environmental control and waste processing facilities of high quality where it cannot be dumped illegally and can only be incinerated at high cost to countries with lower quality environmental control and waste processing. Transport of hazardous waste over long distances is a rather new phenomenon. Sometimes this is done in complete secrecy. In other cases private individuals in the receiving country are paid for the illegal import. In yet other cases the import is accepted by public authorities who are eager to get paid for it, but opposed by citizens_ groups in the receiving country (Vallette and Spalding *et al.*, 1990). World-wide operating environmental organizations like Greenpeace and Friends of the Earth support local citizens in developing countries who do not like having hazardous waste in their back yards (Princen and Finger *et al.*, 1994).

The example of world-wide transport of hazardous waste illustrates the fact that many parties with widely different interests may be involved in global environmental problems. Conflicts of interests are less and less conflicts between 'exporting' and 'importing' countries as a whole but more and more between parties operating internationally, among which multinational firms and non-governmental organizations world-wide operating.

References

Acheson, J.M. (1975) The Lobster Fiefs: Economic and Ecological Effects of Territoriality in the Maine Lobster Industry. *Human Ecology* 3, pp.183-207.

Ajzen, I. (1991) Theory of Planned Behavior. *Organizational Behavior and Decision Processes* 50, pp.179-211.

Ajzen, I. and M. Fishbein (1980) *Understanding Attitudes and Predicting Social Behavior.* Prentice Hall, Englewood Cliffs, NJ.

Albrecht, D., G. Bultena, E. Hoiberg and P. Novak (1982) The New Environmental Paradigm Scale. *Journal of Environmental Education* 13 (3) pp.39-42.

Allen, J., D. Dain and M. Soskin (1993) Using Coupon Incentives in Recycling Aluminum: a Market Approach to Energy Conservation Policy. *Journal of Consumer Affairs* 27 (2) pp.300-318.

Axelrod, R. (1984) *The Evolution of Cooperation.* Basic Books, New York.

Axelsson, A. (1991) Leisure Noise Exposure in Adolescents and Young Adults. *Journal of Sound and Vibration* 151, pp.447-453.

Bachrach, P. and M.S. Baratz (1970) *Power and Poverty. Theory and Practice.* Oxford University Press, New York.

Bacon, F. (1627) *New Atlantis.* Edited by H. Osborne, University Tutorial Press, London.

Baird Callicott, J. (1994) *Earth's Insights. A Multicultural Survey of Ecological Ethics from the Mediterranean Basin to the Australian Outbacks.* University of California Press, Berkeley Ca.

Baird, D.G., R.H. Gertner and R.C. Picker (1995) *Game Theory and the Law.* Harvard University Press, Cambridge, Ma.

Bandura, A. (1982) Self Efficacy Mechanism in Human Agency. *American Psychologist* 37, pp.122-147.

Barker, M.L. (1976) Planning for Environmental Indices: Observers Appraisals of Air Quality. in: K. Craik and E. Zube (eds.) *Perceiving Environmental Quality: Research and Applications.* Plenum Press, New York. pp.172-203.

Barnes, N. (1996) Conflicts over Biodiversity. In: P.B.Sloep and A.Blowers (eds). pp.217-241

Batenburg, R.S. and W.P. Knulst (1993) *Sociaal-culturele Beweegredenen.* Social and Cultural Planning Agency, Rijswijk, the Netherlands.

Baum, A., R. Fleming and J.E. Singer (1982) Stress at Three Mile Island. Applying psychological impact analysis. in L. Bickman (ed.) *Applied Social Psychological Annual Vol.3.* pp.217-248. Sage, Beverly Hills, Ca.

Baum, A., R. Fleming and J.E. Singer (1983) Coping with Victimization by Technological Disaster. *Journal of Social Issues* 39, pp.117-138.

Beck, U. (1992) *Risk Society. Towards a new modern society.* Sage, London.

Beck, U. (1996) Risk Society and the Provident State. in: S. Lash, B. Szerszynsi, B. Wynne (eds.) *Risk, Environment and Modernity.* Sage, London, pp.27-43.

Behnke, R. and C.Kerven (1994) Redesigning for Risk: Tracking and Buffering Environmental Variability in Africa's Rangelands. *Natural Resource Perspectives* 1, Overseas Development Institute, London, November 1994, pp.1-4.

Benedick, R.E. (1991a) *Ozone Diplomacy. New Directions in Safeguarding the Planet.* Harvard University Press. Cambridge, Mass.

Benedick, R.E. (1991b) The Diplomacy of Climate Change. Lessons from the Montreal Ozone Protocol. *Energy Policy* 19 (2) pp.94-97.

Benedick, R.E. and R. Pronovo III (1992) Atmosphere and Outer Space. In: P.H.Sand (ed.) pp.123-148

Benedict, B (1970) Population Regulation in Primitive Societies. In: A.Allison (ed.) *Population Control.* Penguin, Harmondsworth, pp.165-180.

Bennett, G. (1992) *Dilemma. Coping with Environmental Problems.* Earthscan, London.

Benzler, G., M. Halstrick-Schwenk, P. Klemmer and K. Löbbe (1995) *Wettbewerbskonformität von Rücknameverpflichtungen im Abfallbereich.* Rheinisch-Westfählichen Instituts für Wirtschaftsforschung, Essen, Germany.

Berglund, B. and T. Lindvall (1995) *Community Noise. Document Prepared for the World Health Organization.* Center for Sensory Research, Stockholm University & Karolinska Institute, Stockholm, Sweden.

Bernardini, O. and R. Galli (1993) Dematerialization: Long Term Trends in the Intensity of use of Materials and Energy. *Futures* 1993 (May) pp.431-448.

Bilsky (ed.), L.J. (1980) *Historical Ecology. Essays on Environment and Social Change.* National University Publications, Kennikat Press, Port Washington NY/London.

Binswanger, H.C. (1990) Neue Dimensionen des Risikos. *Zeitschrift für Umweltpolitik & Umweltrecht* 13 (2) pp.103-118.

Binswanger, M. (1992) Information und Entropie. *Ökologische Perspektiven des Übergangs zu einer Informationswirtschaft.* Campus, Frankfurt am Main, Germany.

Biesiot, W., H.C. Moll, K. Vringer, H.C. Wilting, K. Blok, R. Kok, K.J. Noorman, J. Potting (1995) *Reduction of Co2 emissions by lifestyle changes.* IVEM; University of Groningen, The Netherlands.

Blackmore, S. (1996) Knowing the Earth's Biodiversity: Challenge for the Infrastructure of Systematic Biology. *Science* 274, pp.63-64.

Blaikie, N.W.H. (1992) The Nature and Origins of Ecological Word Views: An Australian Study. *Social Science Quarterly,* 73, pp.144-165.

Blowers, A.T. and P. Leroy (1994) Power, Politics and Envirpnmental Inequality: a Theoretical and Empirical Analysis of the Process of 'Peripherilization'. *Environmental Politics* 3 (2) pp.197-228.

Blowers, A.T., D. Lowry and B.D. Solomon (1991) *The International Politics of Nuclear Waste.* Macmillan, London.

Boehmer-Christiansen, S. (1994) Politics and Environmental Management. *Journal of Environmental Planning and Management* 37 (1) 69-85.

Boehmer-Christiansen, S. and J. Skea (1991) *Acid Politics. Environmental and Energy Policies in Britain and Germany.* Belhaven Press, London.

Bookchin, M. (1980) *Toward an Ecological Society.* Black Rose Books, Montréal.

Bookchin, M. (1982) *The Ecology of Freedom.* Cheshire Books, Palo Alto, Ca.

Boserup, E. (1965) *The Conditions of Agricultural Growth. The Economics of Agrarian Change under Population Pressure.* George Allen & Unwin Ltd. London.

Boulding, K.E. (1968) General Systems Theory. The Skeleton of Science. In: W.Buckley (ed.) *Modern Systems Research for the Behavioral Scientist.* Aldine, Chicago. pp.3-10.

Bramwell, A. (1989) *Ecology in the 20th Century. A History.* Yale University Press, New Haven, London.

Brehm, J.W. (1966) *A Theory of Psychological Reactance.* Academic Press, New York.

Brenton, T. (1994) *The Greening of Machiavelli. The Evolution of International Environmental Politics.* Earthscan, London.

Brobst, D.B. (1979) Fundamental Concepts for the Analysis of Resource Availability. In: Kerry Smith (ed) *Scarcity and Growth Reconsidered.* The John Hopkins University Press, Baltimore. pp.106-142.

Brockner, J. and J.Z. Rubin (1985) *Entrapment in Escalating Conficts.* Springer, New York.

Burdge, R.J. (1987) The Social Impact Assessment Model and the Planning Process. *Environmental Impact Assessment Review* 7 (2) pp.140-150.

Buttel F.H. (1986) Sociology and the Environment: the Winding Road toward Human Ecology. *International Social Science Journal* 109, pp.337-356.

Buttel F.H. and O.W. Larson (1980) Wither Environmentalism? The Future Political Path of the Environmental Movement. *Natural Resources Journal* 20 (2) pp.323-344.

Buttel, F.H. and W.L. Flinn (1974) The Structure of Support for the Environmental Movement 1968-1970. *Rural Sociology* 39 (1) pp.56-69.

Carraro, C. and J.A. Filar (1995) *Control and Game. Theoretic Models of the Environment.* Birkhauser, Boston.

Carson, R. (1962) *Silent Spring.* Houghton-Mifflin, Boston.

Catton, W.R and R.E. Dunlap (1978) Environmental Sociology: a New Paradigm. *The American Sociologist* 13 (1) pp.41-49.

Catton, W.R and R.E. Dunlap (1980) A New Ecological Paradigm for Post-exurbant Sociology. *American Journal of Behavior Science* 24 (1) pp.15-47.

Chernobyl Conference (1992). "Chernobyl Conference" at the Hebrew University (1992). June 4, 1992. *Environmental Policy Review* Vol.6. pp.21-22.

Cleaver, K.M. and G.A.Schreiber (1994) *Reversing the Spiral. The Population, Agriculture and Environment Nexus in Sub-Saharan Africa.* The World Bank, Washington DC.

Coase, R. (1960) The Problem of Social Cost. *Journal of Law and Economics* 3, pp.1-44.

Cobb, R.W. and C.D. Elder (1971) The Politics of Agenda Building. An Alternative Perspective for Modern Democratic Theory. *The Journal of Politics* 33, pp.892-915.

Cobb, R.W., J.K. Ross and M.H. Ross (1976) Agenda Building as a Comparative Political Process. *American Political Science Review* 70, pp.126-138.

Colborn, T., D. Dumanoski and J.P. Meyers (1996) *Our Stolen Future. Are we threatening our Fertility, Intelligence, and Survival? A Scientific Detective Story.* Plume Penguin, New York.

Commoner, B. (1971) *The Closing Circle. Nature, Man and Technology.* A.A.Knopf, New York.

Costanza, R., R. d'Arge and R.de Groot et al. (1997) The Value of the World's Ecosystem Services and Natural Capital. *Nature* 387 pp.253-260.

Cotgrove, S. (1982) *Catastrophy or Cornucopia. The Environment, Politics and the Future.* Wiley, Chichester.

Couto Soares, M.C. (ed.) (1992) *Debt Swaps, Development and Environment.* IBASE, Rio de Janeiro.

Covello, V.T. (1983) The Perception of Technological Risks: A Literature Review. *Technological Forecasting and Social Change* 23, pp.285-297.

Covello, V.T. and B.B. Johnson (1987) *The Social and Cultural Construction of Risk.* Kluwer, Dordrecht, the Netherlands.

Cox, S.J.B. (1985) No tragedy on the commons. *Environmental Ethics* 7 (1) pp.49-61.

Cramer, J. and J.Schot (1993) Environmental Comakership among Firms as a Cornerstone in the Striving for Sustainable Development. In: Fischer, K. and J.Schot (eds., 1993) pp.311-328

Crenson, M.A. (1971) *The Un-politics of Air Pollution.* The John Hopkins University Press, Baltimore, Md.

Crutzen, P.J. (1971) Ozone Production Rates in an Oxygen-hydrogen-nitrogen Oxide Atmosphere. *Journal of Geophysical Research* 76, pp.7311-7327.

Dahl, R.A. (1961) *Who governs?* Yale University Press, New Haven, London.

Daly, H. (1996) *Free Trade, Capital Mobility and Growth versus Environment and Community.* Public Lecture, Institute of Social Studies, The Hague, The Netherland.

Dawes, R.M., J. McTavish and H. Shaklee (1977) Behavior, Communication, and Assumptions about other People's Behavior in a Commons Dilemma Situation. *Journal of Personality and Social Psychology* 35 (1) pp.1-11.

de Bruyn, S.M. and J.B.Opschoor (1997) Developments in the Throughput-Income Relationship: Theoretical and Empirical Observations. In: *Ecological Economics.* (in press)

de Jasay, A. (1989) *Social Contract Free Ride: a Study of the Public Goods Problem.* Clarendon Press, Oxford.

de Jong, P. and M. Wolsink: The Structure of the Dutch Waste Sector and Impediments for Waste Reduction. *Waste Management & Research* 15 (5) 1997.

de Swaan, A. (1987) *In the Care of the State; Health Care , Education and Welfare in Europa and the USA in the Modern Era.* Polity Press, Cambridge.

Descartes, R. (1965) *Disourse on Method, Optics, Geometry, and Meteorology.* The Bobbs-Merrill Company, Indianapolis (Originally published in 1637).

Devall, B and G. Sessions (1985) *Deep Ecology: Living as if Nature Mattered.* Gibbs M. Smith, Pelegrine Smith Books. Salt Lake City.

Dickson, D. (1974) *Alternative Technology and the Politics of Technical Change.* Glasgow. Fontana/Collins.

Dickson, L. and A. McCulloch (1996) Shell, the Brent Spar and Greenpeace: a Doomed Tryst? *Society and Natural Resources* 9 (1) pp.122-129.

Diekmann, A. (1985) Volunteer's Dilemma. *Journal of Conflict Resolution* 29, pp.605-610.

Dieleman, H., R. van Berkel, F. Reijenga, S. de Hoo, H.Brezet, J.Cramer, J.Schot (1991) PRISMA: Choosing for Prevention is winning. In: *Prepare Experiences.* Ministry of Economic Affairs. The Netherlands.

Dietz, T. (1996) *Entitlements to Natural Resources. Contours of Political Environmental Geography.* International Books, Utrecht, the Netherlands.

Dijksterhuis, E.J. (1969) *The Mechanization of the World Picture.* Oxford University Press, London.

Dillman, D.A., E.A. Rosa and J.J. Dillman (1983) Lifestyle and Home Energy Conservation in the United States. *Journal of Economic Psychology* 3 pp.299-315.

Dinkelman, G. (1995) *Verzuring en Broeikaseffect. De Wisselwerking tussen Problemen en Oplossingen in het Nederlandse Luchtverontreinigingsbeleid. 1970-1994.* Jan van Arkel, Utrecht, the Netherlands.

Dolatyar, M. (1995) Water Diplomacy in the Middle East. In: E.Watkins (ed.) *The Middle Eastern Environment.* St.Malo Press, Cambridge UK, pp.35-43.

Douglas, M. (1982) *Essays on the Sociology of Perception.* Routledge and Kegan Paul, London.

Douglas, M. and A. Wildavsky (1982) *Risk and Culture. An Essay on the Selection of Technical and Environmental Dangers.* University of California Press, Berkeley Ca.

Downs, A. (1972) Up and down with Ecology – The Issue Attention Cycle. *The Public Interest* 2 pp.38-50.

Dubos, R. (1974) Franciscan Conservation versus Benedictine Stewardship. In: D. and E.Spring (eds.). pp.114-136. (Originally published in 1972).

Dunlap, R. and K.D. van Liere (1978) The New Environmental Paradigm: A proposed Measuring Instrument and Preliminary Results. *Journal of Environmental Education* 9, pp.10-19.

Dunlap, R. and K.D. van Liere (1984) Commitment to the Dominant Social Paradigm and Concern for Environmental Quality. *Social Science Quarterly* 4, pp.1013-1028.

Dunlap, R., G.H. Gallup Jr and A.M. Gallup (1993) Of global Concern: Results of the Health of the Planet Survey. *Environment* 35 (9) pp.7-39.

Durkheim, E. (1993) *The Division of Labor in Society.* transl. by G. Simpson, The Mac Millan Company, New York. (originally published 1893).

Eagly, A.H. and S. Chaiken (1993) *Attitudes.* Harcourt Brace Jovanovich College Publishers, Fort Worth, Tex.

Easton, D. (1965) *A Systems Analysis of Political Life.* John Wiley & Sons, New York.

Eberg, J. (1997) *Waste Policy and Learning. Policy Dynamics of Waste Management and Waste Incinration in the Netherlands and Bavaria.* University of Amsterdam, the Netherlands.

Ecologist, The (1972) A Blueprint for Survival. Penguin, Harmondsworth.

Ecologist, The (1988) Special "Deep Ecology" Issue. Vol.18 (4/5).

Eder, K. (1996) The Institutionalisation of Environmentalism: Ecological Discourse and the Second Transformation of the Public Sphere. In: S. Lash, B. Szerszynsi, B. Wynne (eds.) *Risk, Environment and Modernity.* Sage, London. pp.203-224.

Ehrenfeld, J.R., E.P.Craig and J. Nash (1989) Waste Incineration: Confronting the Sources of Disagreement. *Environmental Impact Assessment Review* 9 (3) pp.305-315.

Ehrlich, P. and R. Ehrlich (1970) *Population, Resources, Environment.* Freeman and Co., San Francisco, Ca.

Ehrlich, P. and A. Ehrlich (1990) *The Population Explosion.* Simon and Schuster, New York.

Eiser, J.R., J van der Pligt and R. Spears (1995) *Nuclear Neighbourhoods. Community Responses to Reactor Siting.* University of Exeter Press, Exeter.

Elias, N. (1978) The Civilizing Process. Sociogenetic and Psychogenetic Investigations. Urizen Books. New York. (First German edition in 1939).

EPA, Environment Protection Agency (1992) *Environmental Equity: Reducing the Risk for all Communities.* US EPA, Washington DC.

Erickson, P.A. (1994) *A Practical Guide to Environmental Impact Assessment.* Academic Press, San Diego Ca.

Ester, P. and T. Mandemaker (1994) Socialization of Environmental Policy Objectives. In: Dutch Committee for Long-Term Environmental Policy: *The Environment: Towards a Sustainable Future.* Kluwer, Dordrecht. pp.49-80.

Ester, P., C. Mindell, J. van der Linden and J. van der Pligt (1983) The Influence of Living near a Nuclear Power Plant on Beliefs about Nuclear Energy. *Zeitschrift für Umweltpolitik* 4 (1983) pp.349-386.

Evans, G.W. and R. Tafalla (1987) Measurement of Environmental Annoyance. In Koelega (1987).

Eyre, N. (1994) Externalities of Fuel Cycles, report no.7 *Wind Fuel Cycle, Estimation of Physical Impacts and Monetary Valuation for Priority Impact Pathways.* European Commission DG XII, Brussels, Belgium.

Femers, S. and H. Jungermann (1992) Eine Systematisierumg und Diskussion von Risikomaßen. *Zeitschrift für Umweltkunde* 1, pp.59-84.

Festinger, L. (1957) *A Theory of Cognitive Dissonance.* Harper and Row, New York.

Finsterbusch, K. (1985) State of the Art in Social Impact Assessment. *Environment and Behaviour* 17 (2) pp.193-221.

Fischer, K and J.Schot (1993) *Environmental Strategies for Industry. International Perspectives on Research Needs and Policy Implications.* Island Press, Washington DC.

Fischhoff, B., O. Svensson and P. Slovic (1987) Active Responses to Environmental Hazards: Perception and Decision Making. In D.Stokols and I.Altman (eds.) *Handbook of Environmental Psychology Vol.II.* John Wiley & Sons, New York, pp.1089-1133.

Fischhoff, B., S. Lichtenstein, P. Slovic, S.L. Derby and R.L. Keeney (1981) *Acceptable Risk.* Cambridge University Press, Cambridge Ma.

Fishbein, B.K. (1994) *Germany, Garbage, and the Green Dot. Challenging the Throwaway Society.* Inform, New York.

Fishbein, M. and I. Ajzen (1975) *Belief, Attitude, Intention and Behaviour : an Introduction to Theory and Research.* Addison-Wesley, Reading, Ma.

Flavin, C. and A.B.Durning (1988) *Building on Success: the Age of Energy Efficiency.* Worldwatch Paper 82, Washington DC.

Flavin, C. and N. Lenssen (1994) Reshaping the Electric Power Industry. *Energy Policy* 22 (12) pp.1029-1044.

Freudenburg, W.R. (1987) Perceived Risk, Real Risk: Social Science and the Art of Probabilistic Risk Assessment. *Science* 242, pp.44-49.

Freudenburg, W.R. and S.K. Pastor (1992) NIMBYS and LULUS: Stalking the Syndromes. *Journal of Social Issues* 48 (4) 39-61.

Gandy, M. (1994) *Recycling and the Politics of Urban Waste.* Earthscan, London.

Gijswijt, A.J. (1996) The Establishment of an International Regime: the Case of Stratospheric Ozone Depletion. In: P.Ester and W.Schluchter (eds.) *Social Dimensions of Contemporary Environmental Issues: International Perspectives.* Tilburg University Press. Tilburg, the Netherlands. pp.55-80.

Gipe, P. (1995) *Wind Energy comes of Age.* John Wiley & Sons, New York.

Glacken, C.J. (1967) *Traces on the Rhodian Shore. Nature and Culture in Western Thought from Ancient Times to the End of the Eighteenth Century.* University of California Press. Berkeley Ca.

Goodin, R.E. (1996) Enfranchising the Earth, and its Alternatives. *Political Studies* 54, pp.835-849.

Goodman, P. (1970) *New Reformation. Notes of a Neolithic Conservative.* Random House, New York.

Goodman, P.(without date) *Utopian essays and Practical Proposals.* Vintage, New York.

Gore, A. (1992) *Earth in the Balance. Forging a New Common Purpose.* Earthscan, London.

Gottlieb, R.S. (1996) *This Sacred Earth. Religion, Nature, Environment.* Routledge, New York.

Goudsblom, J. (1992) *Fire and Civilization.* Allen Lane, London.

Grant, W. (1995) *Autos, Smog, and Pollution Control.* Edward Elgar, Aldershot.

Greenberg, M. and D. Schneider (1996) *Environmentally Devastated Neighbourhoods: Perceptions, Realities, and Policies.* Rutgers University Press, New Brunswick.

Gregory, R. and H. Kunreuther (1990) Successful Siting Incentives. *Civil Engineering,* 1990 (4) 73-75.

Groenewegen, P. (ed.) (1996) *The Greening of Industry. Resource Guide and Bibliography.* Island Press, Washington DC.

Grove, R.H. (1995) *Green Imperialism. Colonial Expansion, Tropical Island Edens and the Origins of Environmentalism, 1600-1860.* Cambridge University Press, Cambridge Ma.

Grupp, H. (1986) Die soziale Kosten des Verkehrs, I & II. *Verkehr und Technik* 9 & 10 pp.359-366 & 403-407.

Hajer, M.A. (1995) *The Politics of Environmental Discourse: Ecological Modernisation and the Policy Process*. Clarendon Press, Oxford.

Hajer, M.A. (1996) Ecological Modernisation as Cultural Politics. In: S. Lash, B. Szerszynsi, B. Wynne (eds.) *Risk, Environment and Modernity*. Sage, London, pp.246-268.

Hall, J.A. (1995) *Civil Society. Theory, History, Comparison*. Polity Press, Cambridge UK.

Hamilton, A. (1973) *The Appeal of Fascism. A Study of Intellectuals and Fascism 1919-1945*. Avon Books, New York.

Hannah, L. et al. (1994) A Preliminary Inventory of Human Disturbance of World Ecosystems. *Ambio* 23(4/5) 246-250.

Hannigan, J.A. (1995) *Environmental Sociology. A Social Constructionist Perspective*. Routledge, London, New York.

Hardin, G. (1968) The Tragedy of the Commons. *Science* 162, pp.1243-1248.

Harris, C.C. and G. Brown (1992) Gain, Loss and Personal Responsibility: the Role of Motivation in Resource Valuation Decision-making. *Ecological Economics* 5 (1) pp.73-92.

Harris, N. (1990) *The End of the Third World. Newly Industrializing Countries and the Decline of an Ideology*. Penguin, London.

Hayes, D. (1976) *Energy: the Case for Conservation*. Worldwatch Institute, Washington DC.

Hays, S.P. (1987) *Beauty, Health and Permanence. Environmental Politics in the United States, 1955-1985*. Cambridge University Press, Cambridge, Ma.

Heaton, G.R. and J. Maxwell (1984) Patterns of Automobile Regulation: an International Comparison. *Zeitschrift für Umweltpolitik* 7 (1) pp.15-40.

Heilbroner, R. (1993) *The Making of Economic Society*. 9th ed. Prentice-Hall, Englewood Cliffs, NJ.

Hertz, B. and E. Tellegen (1995) Communication Strategies of the Environmental Movement in the Netherlands. In: W. Leal Filho, F. MacDermott & Z.Murphy (eds.) *Practices in Environmental Education in Europe*. European Research and Training Centre on Environmental Education, Bradford. UK. 85-96

Hirsch, F. (1977) *Social Limits to Growth*. Routledge and Kegan Paul, London.

Hobbes, T. (1971) *Leviathan*. London (first published 1651).

Hobsbawm, E.J. (1979) *Industry and Empire. From 1750 to the Present Day. The Pelican Economic History of Britain. Volume 3*. Penguin, Harmondsworth.

Hohmeyer, O. (1988) *The Social Costs of Energy Consumption. External Effects of Electricity Generation in the Federal Republic of Germany*. Springer, Berlin, Germany.

Hohmeyer, O. (1992) Renewables and the Full Cost of Energy. *Energy Policy* 20 (4) pp.365-375.

Holy Bible (1992) New International Version. International Bible Society, Pocket cross Reference edition. Hodder & Stoughton, London.

Homer-Dixon, T.F. (1994) Environmental Scarcities and Violent Conflicts. Evidence form Case. *International Security* 19, pp.5-40.

Hopper, J.R. and J.M. Nielsen (1991) Recycling as Altruistic Behavior: Normative and Behavioral Strategies to expand Participation in a Community Recycling Program. *Environment and Behavior* 23 (2) 195-220.

Hordijk a.o.; L. (1991) Use of RAINS Model in Acid Rain Negotiations in Europe. *Environmental Science and Technology*. 25 (4) 596-602.

Huber, J. (1982) *Die verlorene Unschuld der Oekologie. Neue Technologien und superindustrielle Entwicklung*. Fischer Verlag, Frankfurt am Main, Germany.

Huckestein, B. (1996) Ökologische Steuerreform und Nachhältige Entwicklung; Ansatzpunkte und Bestandteile einer Nachhaltigen Finanzreform. *Zeitschrift für Umweltpolitik & Umweltrecht* 19 (3) pp.387-408.

Hueting, R. (1980) *New Scarcity and Economic Growth. More Welfare through less Production?* North Holland Publishing Company. Amsterdam, New York, Oxford.

Humphrey, C.R. and F.R. Buttel (1982) *Environment, Energy, and Society.* Wadsworth, Belmont CA.

Hurrell, A (1992) Brazil and the International Politics of Amazonian Deforestation. In: Hurrell, A. and B.Kingsbury (eds.) *The International Politics of the Environment. Actors, Interests and Institutions.* Clarendon Press, Oxford. pp. 398-429.

IEA/ENEL (1991) *Conference on Advanced Technologies for Electric Demand-side Management. Preliminary Proceedings.* Sorrento.

Inhaber, H. (1997) *Slaying the Nimby Dragon.*

IPPC (1992) *The Supplementary Report to the IPPC Scientific Assessment of Climate Change 1992.* Houghton, J.T., Callander, B.A. and Varney, S.K. (Eds). Cambridge University Press, Cambridge.

IUCN (1994) Netherlands Committee for IUCN - *The Netherlands and the World Ecology* (publisher not mentioned).

Jacobson, T. and R.M. Adams (1958) Salt and Silt in Ancient Mesopotamian Agriculture. *Science* 1958, pp.1251-1258.

Jacobsson, L. (1991) From Project to Programme: commercializing Efficient Electricity Use to Reduce Power System Costs. Vattenfall, Sweden. In: IEA/ENEL, 1991.

Jänicke, M. (1990) Erfolgsbedingungen von Umweltpolitik im Internationalen Vergleich. *Zeitschrift für Umweltpolitik & Umweltrecht* 13 (3) pp.213-232.

Jänicke, M., H. Mönch and M. Binder (1993) Ecological Aspects of Structural Change. Freie Universität Berlin, Berlin.

Jänicke, M., and H. Weidner (1995) *Successful Environmental Policy: a Critical Evaluation of 24 Cases.* Ed. Sigma, Berlin.

Jänicke, M., H. Mönch, T. Rannenberg and U.E. Simonis (1988) *Economic Structure and Environmental Impact: Empirical Evidence on Thirty-one Countries in East and West.* Science Center, Berlin.

Janis, I.L. and L. Mann (1977) *Decision Making: a Psycho-logical Analysis of Conflict, Choice and Commitment.* The Free Press, New York.

Jasanoff, S. (1992) Science, Politics, and the Negotiation of Expertise at EPA. *Osiris* 7, pp.1-23.

Jenkins-Smith, H.C. (1991) Alternative Theories of the Policy Process: Reflections on Research Strategy for the Study of Nuclear Waste Policy. *PS; Political Science and Politics* 24, pp.157-166.

Job, R.F.S. (1988) Community Response to Noise: A Review of Factors Influencing the Relationship between Noise Exposure and Reaction. *Journal of the Accoustical Society of America* 83, pp.991-1001.

Job, R.F.S. (1989) Community Reaction to Noise from Power Stations. *Internoise 89*, Newport Beach CA, pp.865-868.

Johansen, S. and D.T. Hoog (1995) European Benefit/Cost (EUBC) Analysis Methodology for Energy Efficiency Programs. In A. Persson (ed.) *Sustainability and the Reinvention of Government* ECEEE, Nutek, Stockholm, Sweden.

Kahneman, D. and A. Tversky (1984) Choices, Values, and Frames. *American Psychologist* 39 (4) pp. 341-350.

Kahneman, D. and A.Tversky (1979) Prospect Theory: an Analysis of Decision under Risk. *Econometrica* 47, pp 263-291.

Kahneman, D., P. Slovic and A. Tversky (1982) *Judgment under Uncertainty: Heuristics and Biases*. Cambridge University Press, Cambridge Ma.

Kastka, J. (1982) *Erfassung und Bewertung von Geruchen und ihren Belästigungswirkung*. Universität Wiesbaden, Germany.

Keeney, R.L. and H.Raiffa (1976) *Decisions with Multiple Objectives: Preferences and Value Trade-offs*. John Wiley & Sons, New York.

Kempton, W., J.S. Boster and J.A. Hartley (1992) *Environmental Values in American Culture*. MIT Press, Cambridge, Ma.

Kingdon, J.W. (1995) *Agenda's, Alternatives and Public Policies*. 2nd ed. Harper Collins, New York.

Kinnersley, D. (1988) Troubled Waters. *Rivers, Politics and Pollution*. London.

Kleinhesselink, R. and E.A. Rosa (1991) Cognitive Representation of Risk Perceptions. *Journal of Cross-Cultural Psychology* 22 (1) pp.11-28.

Kloos, P. (1978) *Evolutie*. Delftse Universitaire Pers, Delft, the Netherlands.

Koelega, H.S. (1987) ed., *Environmental Annoyance: Characterization, Measurement, and Control*. Elsevier, Amsterdam, the Netherlands.

Kolk, A (1996) *Forests in International Politics. International Organisations, NGOs and the Brazilian Amazon*. Thesis University of Amsterdam.

Kriesi, H., R.Koopmans, J.W.Duyvendak and M.G.Giugni (1995) *New Social Movements in Western Europe. A Comparative Analysis*. UCL Press, London.

Kroeze, C. and L. Reijnders (1992) Halocarbons and Global Warming I, II & III *The Science of the Total Environment* 111, pp.1-24; & 112, pp.269-314.

Kropotkin, P. (1902) *Mutual Aid. A Factor of Evolution*. Extending Horizons Books, Boston Ma.

Kryter, K.D. (1985) *The Effect of Noise on Man*. 2nd ed. Academic Press, New York.

Langeweg, F (ed.) (1989) *Concern for Tomorrow. National Institute of Public Health and Environmental Protection*. RIVM, Bilthoven, the Netherlands.

Lefcourt, H.M. (1982) *Locus of Control: Current Trends in Theory and Research* 2nd ed. Lawrence Erlbaum, Hillsdale NJ.

Lehmann, P.M. (1992) Deforestation and Changing Land-use Patterns in Costa Rica. in: H.K Steen and R.P. Tucker (eds.) *Changing Tropical Forests. Historical Perspectives on today's Challenges in Central & South America*. CEDLA, pp.58 -76.

Lenski, G. (1966) *Power and Privilege. The Theory of Social Stratification*. McGraw-Hill, New York.

Levine, A.G. (1982) *Love Canal: Science, Politics, and People*. Lexington Books, Lexington.

Lichtenstein, S., P. Slovic, B. Fischhoff, M. Layman and B. Combs (1978) Judged Frequency of Lethal Events. *Journal of Experimental Psychology, Human Learning and Memory* 4, pp.551-579.

Lindblom, C.E. (1977) *Politics and Markets. The Worlds Political Economic Systems*. Basic Books, New York.

Lovelock, J.E. (1988) *Gaia. A New Look at Life on Earth*. Oxford University Press, Oxford.

Luce, R.D. and H. Raiffa (1957) *Games and Decisions: Introduction and Critical Survey*. John Wiley & Sons, London.

Lucretius (1951) *On the Nature of the Universe*. Translated and introduced by R.E.Latham. Penguin, Harmondsworth.

Maccy, S.M. and M.A. Brown (1983) Residential Energy Conservation. The Role of Past Experience in Repetitive Household Behavior. *Environment and Behavior* 15 (2) 123-141.

Mackerron, G. (1989) The Decommissioning of Nuclear Plants: Timing, Cost and Regulation. *Energy Policy* 17 (2) pp.103-108.

Maloney, M.P. and M.P. Ward (1973) Ecology, let's hear it from the People. An Objective Scale for the Measurement of the Ecological Attitudes and Knowledge. *Americal Psychologist* 28, pp.583-586.

Malthus, T.R. (1976) *An Essay on the Principle of Population and a Summary View of the Principle of Population.* Penguin, Harmondsworth (first published 1798).

Manstead, A.S.R. and D. Parker (1995) Evaluating and Extending the Theory of Planned Behaviour. *European Review of Social Psychology* 6, pp.71-95.

Mansvelt Beck, J. (1981) *Milieu en Derde Wereld.* Delft University Press. Delft, the Netherlands.

Marks, G. and D. von Winterfeldt (1984) "Not in My Back Yard": Influence of Motivational Concerns on Judgments about Risky Technology. *Journal of Applied Psychology* 69, pp.408-415.

Marsh, G.P. (1864) *Man and Nature; or, Physical Geography as modified by Human Action.* Charles Schribner, New York, 1864. (Re-edited: The Belknap Press of Harvard University Press. Cambridge, Ma., 1965).

Marx, K. (1976) *Capital. A Critique of Political Economy. Volume One.* Penguin, Harmondsworth, (first German edition 1867).

Maslow, A.H. (1943) A Theory of Human Motivation. *Psychological Review* 50, pp.370.

Maslow, A.H. (1971) *The Farther Reaches of Human Nature.* Viking, New York.

Maull, H.W. (1992) Japan's Global Environmental Policies. In: Hurrell and Kingsbury (eds.) pp.354-372.

McGowan, F. (1991) Controlling the Greenhouse Effect. The Role of Renewables. *Energy Policy* 19 (2) pp.110-118.

McGuire, W.J. (1985) Attitudes and Attitude Change. in: G. Lindzey and E. Aronson (ed.) *Handbook of Social Psychology Vol II Special Fields and Applications.* pp.233-346. Random House, New York.

McKibben,W. (1989) *The End of Nature.* Random House, New York.

McNeill, W.H. (1976) *Plagues and People.* Anchor Press/Doubleday, New York.

Meadows, D.H., D.L. Meadows, J. Randers and W.H. Behrens III (1972) *The Limits to Growth. A Report for the Club of Rome's Project on the Predicament of Mankind.* Universe Books, New York, 1974 (1st ed. 1972).

Medvedev, Z. (1990) *The Legacy of Chernobyl.* Basil Blackwell, Oxford.

Melnick, P.S. (1984) Pollution Deadlines and the Coalition for Failure. *The Public Interest* 75, pp.123-134.

Merchant, C. (1983) *The Death of Nature. Women, Ecology and the Scientific Revolution.* Harper, SanFrancisco Ca.

Milbrath, L.W. (1984) *Environmentalists: Vanguard for a New Society.* State University of New York Press, Albany, NY.

Mill, J.S. (1965) *Collected Works of John Stuart Mill, vol.III.* J.M. Robson (ed.). University of Toronto Press, Toronto, Can.

Mill, J.S. (1970) *Principles of Political Economy.* Penguin, Harmondsworth. (originally published in 1848).

Milton, K. (1996) *Environmentalism and Cultural Theory. Exploring the Role of Anthropology in Environmental Discourse.* Routledge, London.

Mitchell, R.C. and R.T. Carson (1986) *Using Surveys to Value the Benefits of Public Goods: the Contingent Valuation Method.* Resources for the Future, Washington DC.

Mol, A.P.J. (1995) *The Refinement of Production. Ecological Modernization Theory and the Chemical Industry.* Van Arkel. Utrecht, the Netherlands.

Molina, M.J. and F.S. Rowland (1974) Stratospheric Sink for Chlorofluoromethanes. Chlorine Atomic catalysed Destruction of Ozone. *Nature* 249, pp.810-812
Moncrief, L.W. (1974) The Cultural Basis for our Environmental Crisis. In: D.and E.Spring (1974). pp.76-90. (originally published in 1970).
Moore, B (1966) *Social Origins of Dictatorship and Democracy. Lord and Peasant of the Modern World.* Beacon Press, Boston Ma.
Morell, D. and C. Magorian (1982) *Siting Hazardous Waste Facilities.* Ballinger, Cambridge, Ma.
Moyo, S., P. O'Keefe and M. Sill (1993) *The Southern African Environment. Profiles of the SADC Countries.* Earthscan, London.
Muntingh, H. (1995) Natuurbescherming versus Mensenbescherming. De Afweging van een Politicus. *Socialisme en Democratie* 52 (6) 279-284.
Myers, M. and J.L.Simon (1994) *Scarcity or Abundance? A Debate on the Environment.* W.W.Norton & Company, New York/London.
Nadel, S. (1992) Utility Demand Side Management Experience and Potential - a Critical Review. *Annual Review of Energy and the Environment Vol.17*, pp.507-535. Annual Review Inc., Palo Alto Ca.
Navarro, P. (1981) The 1977 Clean Air Act Amendments: Energy, Environmental, Economic, and Distributional Impacts. *Public Policy* 29, pp.121-145.
Neidert, G.P.M. and D.E. Linder (1990) Avoiding Social Traps: some Conditions that Maintain Adherence to Restricted Consumption. *Social Behaviour* 5 (4) pp.261-284.
Nelissen, N. and P. Scheepers (1993) Ecological Conciousness and Behaviour examined. An Empirical Study in the Netherlands. *Sociale Wetenschappen* 35 (4) 64-81.
Nordhaus, W.D. (1991) To slow or not to slow: the Economics of Global Warming. *Economic Journal* 101, pp.920-937.
Noske, B (1989) *Humans and other Animals. Beyond the Boundaries of Anthropology.* Pluto Press, London.
O'Hare, M. (1977) "Not on MY Block you don't": Facility Siting and the Strategic Importance of Compensation. *Public Policy* 25 (4) pp.407-458.
O'Hare, M., L.Bacow and D.Sanderson (1983) *Facility Siting and Public Opposition.* Van Nostrand Reinhold, New York.
O'Sullivan, A. (1993) Voluntary Auctions for Noxious Facilities: Incentives to Participate and the Efficiency of Siting Decisions. *Journal of Environmental Economics and Management* 25 (1) 12-26.
OECD (1991) *The State of the Environment.* OECD, Paris.
OECD (1993) *World Energy Outlook to the Year 2010.* OECD, Paris.
OECD (1995) *Motor Vehicle Pollution. Reduction Strategies beyond 2010.* OECD, Paris.
Öhrström, E. (1991) Sleep Disturbance, Psycho-social and Medical Symptoms - a Pilot Survey among Persons Exposed to High Levels of Road Traffic Noise. *Journal of Sound and Vibration* 133, pp.117-128.
Olson, M. (1971) *The Logic of Collective Action. Public Goods and the Theory of Groups.* (2e ed) Harvard University Press, Cambridge Ma.
Opschoor, J.B. (1996) Sustainability: a Suspicious Concept. In: F. Duijnhouwer and M. Veldhuis (eds.) *Agenda 21: RAWOO/RMNO Lectures on Sustainable Development.* Advisory Council for Scientific Research in Development Problems (RAWOO) The Hague, the Netherlands, pp.8-22.
Ostrom, E. (1986) An Agenda for the Study of Institutions. *Public Choice* 48, 3-25.
Ostrom, E. (1990) *Governing the Commons: the Evolution of Institutions for Collective Action.* Cambridge University Press, Cambridge.
Ostrom, E., R. Gardner and J. Walker (1994) *Rules, Games and Common Pool Reserves.* Michigan University Press, Ann Arbor Mi.

Paehlke, R.C. (1989) *Environmentalism and the Future of Progressive Politics.* Yale University Press, New Haven, London.

Passmore, J. (1974) *Man's Responsability for Nature. Ecological Problems and Western Traditions.* Duckworth, London.

Pasztor, J. (1991) What Role can Nuclear Power Play in Mitigating Global Warming? *Energy Policy* 19 (2) pp.98-109.

Pearce, D. W. and R.K. Turner (1990) *Economics of Natural Resources and the Environment.* Harvester Wheatsheaf, New York.

Perrow, C. (1984) *Normal Accidents. Living with High-risk Technologies.* Basic Books, New York.

Peters, B.G. and B.W. Hogwood (1985) In search of the Issue-attention Cycle. *Journal of Politics* 47, pp.238-253.

Petts, J. (1992) Incineration Risk Perceptions and Public Concern: Experience in the UK improving Risk Communication. *Waste Management & Research* 10 (2) 169-182.

Petty, R.E. and J.T. Cacioppo (1986) *Communication and Persuation. Central and Peripheral Routes to Attitude Change.* Springer, New York.

Pigou, A.C. (1920) *The Economics of Welfare.* Macmillan, London.

Platt, J. (1973) Social Traps. *American Psychologist* 28, pp.641-651.

Ponting, C. (1991) *A Green History of the World.* Penguin Books, Harmondsworth/London.

Popper, F.J. (1985) The Environmentalist and the LULU. *Environment* 27(2) 7-11.

Powell, J. (1993) Thermal Plastics Processing: is it Recycling? *Resource Recycling* May 1993, pp.52-55.

Preisendörfer, P. (1996) Ökologisches Bewußtsein in Ost- und Westdeutschland. *Zeitschrift für Umweltpolitik und Umweltrecht* 19 (1) 1-20.

Princen, T. and M. Finger (1994) *Environmental NGOs in World Politics. Linking the Local and the Global.* Routledge, London and New York.

Prunier, G. (1995) *The Rwanda Crisis. History of a Genocide.* C. Hurst & Co, London.

Rabe, B.G. (1991) Beyond the Ninby-Syndrome in Hazardous Waste Facility Siting. *Governance* 4 (2) 184-206.

Rankin, W,L., B.D. Melber, T.D. Overcast and S.M. Nealey (1981) *Nuclear Power and the Public: an Update of Collected Survey Research on Nuclear Power.* Batelle Memorial Inst., Seattle Wa.

Reijnders, L. (1996a) De Effectiviteit van Mondiaal Milieubeheer. *Amsterdams Sociologisch Tijdschrift* 23 (1) 218-241.

Reijnders, L. (1996b) *Environmentally Improved Production Processes and Products: an Introduction.* Kluwer, Dordrecht, the Netherlands.

Renn, O. (1990) Public Responses after Chernobyl: Effects on Attitudes and Public Policies. *Journal of Environmental Psychology* 10 pp.151-168.

Renn, O., T. Webler and P. Wiedemann (1995) *Fairness and Competence in Citizen Participation.: Evaluating Models for Environmental Discourse.* Kluwer, Dordrecht, the Netherlands.

Roberts, N (1989) *The Holocene. An Environmental History.* Blackwell, Oxford.

Rokeach, M. (1973) *The Nature of Human Values.* Macmillan, New York.

Rosa, E.A. and W.R. Freudenburg (1993) The Historical Development of Public Reactions to Nuclear Power: Implications for Nuclear Waste Policy. in: R.E. Dunlap, M.E. Kraft and E.A. Rosa (eds.) *Public Reactions to Nuclear Waste: Citizens' Views of Repository Siting.* Duke University Press, Durham. pp.33-63.

Rostow, W.W. (1962) *The Stages of Economic Growth. A non-communist Manifesto.* Cambridge, The University Press.

Rousseau, J.J. (1979) *Emile or on Education.* Penguin, London, Harmondsworth (originally published in 1762).
Rowell, A (1996) *Green Backlash. Global Subversion of the Environmental Movement.* Routledge, London/New York.
Rüdig, W. (1990) *Anti-nuclear Movements. A World Survey of Opposition to Nuclear Energy.* Longman Group, Harlow UK.
Sabatier, P.A. (1987) Knowledge, Policy-oriented Learning, and Policy Change. An Advocacy for a Coalition Framework. *Knowledge Creation, Diffusion, Utilization* 8 (4) 649-692.
Sabatier, P.A. and H.C. Jenkins-Smith (1993) *Policy Change and Learning; an Advocacy Coalition Approach.* Westview Press, Boulder, Co.
Sabine, G.H. (1967) *A History of Political Theory.* 3rd ed., Harrap, London.
Sagoff, M. (1984) Ethics and Economics in Environmental Law. in: T. Regan (ed.) *Earthbound.* Random House, New York, pp.147-187.
Sahlins, M. (1972) *Stone Age Economics.* Aldine-Atherton, Chicago, New York.
Samuelson, C.D. (1990) Energy Conservation: a Social Dilemma Approach. *Social Behaviour* 5 (4) 207-230.
Sand, P.H. (1992) *Effectiveness of International Environmental Agreements: Survey of Existing Legal Instruments.* Grotius Publications, Cambridge UK.
Schama, S. (1995) *Landscape and Memory.* Harper Collins, London.
Schattschneider, E.E. (1960) *The Semisovereign People. A Realist's View of Democracy in America.* Holt, Rinehart and Winston, New York.
Schipper, L. (1990) *Energy Use in Passenger Transport in OECD Countries: Changes between 1970 and 1987.* Lawrence Livermore Laboratories, Berkeley, Ca.
Schipper, L. (1996) Life-styles and the Environment: the Case of Energy. *Daedalus* 125 (3) 43-71.
Schipper, L. and S.Meyers (1992) *Energy Efficiency and Human Activity: Past Trends, Future Prospects.* Cambridge University Press, Cambridge UK.
Schipper, L., S. Barlett, D. Hawk and E. Vine (1989) Linking Life-Styles and Energy Use: A Matter of Time. *Annual Review of Energy 1989*, pp.273-320.
Schmidt, O. and R. Van der Veen (1976) Rationaliteit en Moraal. Hit Prisoners' Dilemma. *Acta Politica* 2, pp.178-205.
Schneider, S.H. (1990) Debating GAIA. *Environment* 32 (4) 5-32.
Schultz, T.J. (1978) Synthesis of Social Surveys on Noise Annoyance. *Journal of the Accoustical Society of America* 64, pp.377-405.
Schultz, T.J. (1982) *Community Noise Rating.* Applied Science Publishers, New York.
Schumacher, E.F. (1973) *Small is Beautiful.* Blond & Briggs, London.
Schwarz, M. and M. Thompson (1990) *Divided we stand. Redefining Politics, Technology and Social Choice.* Harvester Wheatsheaf, New York.
Schwartz, S.H. (1977) Normative Influences on Altruism. *Advances in Experimental Social Pschychology* 10, pp.221-279.
Sen, A.K. (1973) Behaviour and the Concept of Preference. *Economica* 151, pp.241-259.
Sen, A.K. (1974) Choice, Orderings and Morality. in: S. Körner (ed.) *Practical Reason.* Blackwell, Oxford, pp.54-67.
Seneca, J.J. and M.K. Taussig (1979) *Environmental Economics.* (2nd ed.) Prentice Hall, Englewood Cliffs NJ.
Shaw, R.P. (1992) The Impact of Population Growth on Environment. *Environmental Impact Assessment Review* 12 (1/2) 11-36.
Sheldrake, R. (1993) *The Rebirth of Nature. New Science and the Revival of Animism.* Rider, London.

Shippee, G., J. Burroughs and S. Wakefield (1980) Dissonance Theory revisited: Perception of Environmental Hazards in Residential Areas. *Environment and Behaviour* 12 (1) pp.33-51.

Simon, H.A. (1987) Rationality in Psychology and Economics. in: Hogarth, R.J. and M.W. Reder (eds.) *Rational Choice: the Contrast between Economics and Psychology.* University of Chicago Press, Chicago Il., pp.25-40.

Simon, J. (1994) Pre-debate Statement. In: Myers, M. and J. Simon: pp.5-65

Sinker, M. (1989) Eco Guerrillas. *The Face*, December 1989, pp.56-58.

Siskind, E. and L.E. Susskind (1989) The Incineration Conflict: Adressing Public Concerns. *Environmental Impact Assessment Review* 9 (3) 317-329.

Sivacek, J. and W.D. Crano (1982) Vested Interest as a Moderator of Attitude - Behavior Consistency. *Journal of Personality and Social Psychology* 43 (2) 210-221.

Slingerland, S. (1997) Energy Conservation and Organization of Electricity Supply in the Netherlands. *Energy Policy* 25 (2) 193-203.

Sloep, B.P. and A. Blowers (ed.) (1996) *Environmental Problems as Conflicts of Interest. Environmental Policy in an International Context. Volume 2.* Open University of the Netherlands & Open University United Kingdom. Arnold, London.

Smil, V. (1994) *China's Environmental Crisis. An Inquiry into the Limits of National Development.* M.E.Sharpe, Armonk, New York, London.

Smith, R.J. (1996) Sustainability and the Rationalisation of the Environment. *Environmental Politics* 5 (1) 25-47.

Sorokin, D. and J. Schulkin (1991) Environmentalism and the Right-to-know: Expanding the Practices of Democracy. *Ecological Economics* 19, pp.316-330.

Spring, D. and E. Spring (1974) *Ecology and Religion in History.* Harper Torchbooks, New York.

Stallings, R. (1990) Media Discourse and the Social Construction of Risk. *Social Problems* 37, pp.80-95.

Stanners, D. and Ph.Bourdeau (eds.) (1995) *European Environment.* The Dobris Assessment, European Environmental Agency, Copenhagen, Denmark.

Starbuck, W.H. (1971) *Organizational Growth and Development.* Penguin, Harmondsworth.

Starr, C. (1969) Social Benefit versus Technological Risk. *Science* 165, pp.1252-1238.

Stern, P.C. (1976) Effects of Incentives and Education on Resource Conservation Decisions in a Simulated Commons Dilemma. *Journal of Personality and Social Psychology* 34 (6) 1285-1292.

Stern, P.C., T. Dietz and J.S. Black (1986) Support for Environmental Protection: the Role of Moral Norms. *Population and the Environment* 8, pp.204-222.

Stern, P.C., T. Dietz and L. Kalof (1992) Value Orientations, Gender, and Environmental Concern. *Environment and Behavior* 25 (3) 322-348.

Stern, P.C. and S. Oskamp (1987) Managing Scarce Environmental Resources. In: Altman and Stokols (eds.) Handbook of Environmental Psychology. Vol. II. John Wiley, New York. pp.1043-1088.

Sugden, R. (1985) Consistent Conjectures and Voluntary Contributions to Public Goods: why the Conventional Theory does not work. *Journal of Public Economics* 27, pp.117-124.

Sutton, D.P. and N.P.Harmon (1973) *Ecology: Selected Concepts.* John Wiley & Sons, New York.

Szagun, G. and E. Mesenholl (1993) Environmental Ethics: an Empirical Study of West-German Pupils. *Journal of Environmental Education* 25 (1) 37-44.

Taylor, D.E. (1989) Blacks and the Environment: Toward an Explanation of the Concern and Action Gap between Blacks and Whites. *Environment and Behavior* 21 (29 175-205.

Taylor, M. and H.Ward (1982) Chickens, Whales and Lumpy Goods: Alternative Models of Public Goods Provision. *Political Studies* 30 (1982) pp.351-370.

Teger, A. (1980) *Too much Invested to quit.* New York.

Tellegen, E (1989) Perestroika and the Rational Use of Materials and Energy. *The Environmental Professional* 11, pp.142-151.

Tellegen, E (1996) Environmental Conflicts in Transforming Economies: Central and Eastern Europe. in: P.B.Sloep and A.Blowers (1996) pp.67-96.

Tellegen, E. and W.Gilijamse (1995) Energy Conservation and the Dutch Energy Sector. *Journal of Economic and Social Geography* 86 (4) pp. 390-395

Tertoolen, G., B. Verstraten and D. van Kreveld (1997) Free to move...?! Psychological Resistance against Attempts to reduce Private Car Use. *Transportation Research* (forthcoming).

The Global Challenge (1995) *"The Global Challenge". Fourth International Energy Efficiency & DSM Conference Berlin.* SRC International Inc., Bala Cynwyd, Penn.

Thogersen, J. (1996) Recycling and Morality. A Critical Review of the Literature. *Environment and Behavior* 28 (4) 536-558.

Thomas, K. (1984) *Man and the Natural World. Changing Attitudes in England 1500-1800.* Penguin, London.

Thompson, M., R. Ellis and A. Wildavsky (1990) *Cultural Theory.* Westview, Boulder, Co.

Thurow, L.C. (1980) *The Zero-sum Society.* Penguin, Harmondsworth.

Titmuss, R.M. (1970) *The Gift Relationship. From Human Blood to Social Policy.* Pelican, Harmondsworth.

Tönnies, F. (1963) *Community and Society.* Translated and edited by C. Loomis, Harper Torch Book, New York (first published 1887).

Toynbee, A (1974) The Religious Background of the Present Environmental Crisis. In: D. and E. Spring (1994) pp.137-149. (originally published in 1972).

Tsuru,S. and H.Weidner (1985) *Ein Modell für uns. Die Erfolge der Japanischen Umweltpolitik.* Köln, Germany.

Tversky, A. and D. Kahneman (1973) Availability: a Heuristic for judging Frequency and Probability. *Cognitive Psychology* (1973) 4, pp.207-232. reprinted in: Kahneman, Slovic and Tversky (1982) pp.163-178.

Udo de Haes, H. (1991) Milieukunde. Begripsbepaling en Afbakening. In: J.J.Boersema a.o.(eds.) *Basisboek Milieukunde.* 4th edition. Boom, Meppel, the Netherlands. pp.21-34.

UNEP (1995) *Global Diversity Assessment.* Cambridge University Press, Cambridge UK.

Uusitalo, L. (1986) *Environmental Impacts of Consumption Patterns.* Gower, Aldershot.

Vallette, J. and H. Spalding (eds.) (1990) *The International Trade in Wastes. A Greenpeace Inventory.* Greenpeace USA, Washington DC.

van Berkel, C.W.M. (1996) *Cleaner Production in Practice.* Thesis. University of Amsterdam.

van Berkel, C.W.M. and J.G.M. Kortman (1993) Waste Prevention Assessments in Small and Medium Sized Enterprises. *Journal of Cleaner Production* 1 (1) pp.21-28.

van Beukering, P. and P. Vellinga (1996) Climate Change: From Science to Global Politics. In: P.B.Sloep and A.Blowers (1996) pp.187-215.

van de Braak, H. (1995) *The Prometheus Complex. Man's Obsession with Superior Technology.* Enzo Press, Amersfoort, the Netherlands.

van der Eijk, C., and W.Kok (1975) Non-decisions Reconsidered. *Acta Politica* 10 pp.277-301.

van der Pligt, J., J.R.Eiser and R.Spears (1986) Attitudes towards Nuclear Energy. Familiarity and Salience. *Environment and Behaviour* 18 (1) pp.75-93.

van der Pot, J.H.J. (1985) *Die Bewertung des technischen Fortschritts. Eine systematische Uebersicht der Theorien.* Van Gorcum, Assen, the Netherlands.

Vernadsky, V.I. (1991) *Nauchnaja mysl'kak Planetnoe Javlenie.* (Scientific Thought as planetarian phenomenon), Nauka, Moskva Russian Republic.

Vernadsky, V.I. (1945) The Biosphere and the Noösphere. *American Scientist* 33 (1) 1-12.

Verplanken, B. (1989) *Persuasive Communication of Technological Risks. A Test of the Elaboration Likelihood Model.* Thesis Leiden University, Leiden, the Netherlands.

Verplanken, B., H. Aarts, A. van Knippenberg and C. van Knippenberg (1994) Attitude versus General Habit: Antecedents of Travel Mode Choice. *Journal of Applied Social Psychology* 24, pp.285-300.

Vine, E. (1995) International DSM and DSM Program Evaluation: an INDEEP Assessment. A.Persson (ed.) *Sustainability and the Reinvention of Government.* ECEEE, Nutek, Stockholm, Sweden.

Vining, J. and A. Ebreo (1992) Predicting Recycling Behaviour from Global and Specific Environmental Attitudes and Changes in Reclycling Opportunities. *Journal of Applied Social Psychology* 22, 1580-1607.

Visser, S. (1987) De huidige energie situatie in de wereld. *Elektrotechnick* 65 (12) pp.250.

Vitousek, P.M. a.o. (1986) Human Appropriation of the Products of Photosynthesis. *BioScience* 36 (6) 369-373.

von Neumann, J. and O. Morgenstern (1947) *Theory of Games and Economic Behaviour.* Princeton University Press, Princeton NJ.

von Neumann, J. (1928) Zur Theorie der Gesellschaftspiele. *Mathematische Annalen* 100, pp.295-320.

von Prittwitz, V. (1990) *Das Katastrophen-Paradox. Elemente einer Theorie der Umweltpolitik.* Leske & Budrich, Opladen, Germany.

von Weizsäcker, E.U., A.B. Lovins and L.H. Lovins (1995) *Faktor Vier. Doppelter Wohlstand, Halbierter Naturverbrauch.* Droener Knaus, München. English version: Factor Four. Doubling Wealth – Halving Resource Use. Earthscan, London, 1997.

von Weizsäcker, E.U., and J. Jesinghaus (1992) *Ecological Tax Reform: a Policy Proposal for Sustainable Development.* Earthscan, London.

Vringer, K. and K. Blok (1995) The Direct and Indirect Energy Requirement of Households in the Netherlands. *Energy Policy* 23 (10) 893-910.

Wachs, T.D. and G.E.Gruen (1982) *Early Experience and Human Development.* Plenum Press, New York

Wahba, M.A. and L.G. Bridwell (1976) Maslow reconsidered: a Review of Research of the Need Hierarchy Theory. *Organizational Behavior and Human Performance.* 15, pp.212.

Warnick, I.K., R. Herman, S. Govind, and J.H. Ausubel (1996) Meterialization and Dematerialization: Measures and Trends. *Daedalus* 125 (3) 171-198.

WCED (1987) World Commission on Environment and Development (WCED) *Our Common Future* (Brundtland Report). Oxford University Press, Oxford/New York.

Weale, A. (1992) *The New Politics of Pollution.* Manchester University Press, Manchester UK.

Weber, M. (1919) *Max Weber's "Science as a Vocation".* Edited by P.Lassman, J.Velody with H,Martins. Unwin Hyman, London, 1989. (Text of a speech held in 1919).

Weber, M. (1968) *Economy and Society. An Outline of Interpretive Sociology.* Edited by G. Roth and C. Wittich. Bedminster Press, New York (first published 1922).

White, L. (1974) The Historical Roots of our Ecological Crisis. In D. and E.Spring (1974) pp.15-31 (originally published in 1967).

Wilkinson, R.G., (1973) *Poverty and Progress.* Mcthuen & Co., London.

Williams, H.E., J. Medhurst and K. Drew (1993) Corporate Strategies for a Sustainable Future. In: Fischer and Schot (1993) pp.117-146.

Williamson, O. (1979) Transaction-cost Economics: the Governance of Contractual Relations. *Journal of Law and Economics* 22 (2) pp.233-261.

Wilson, D.C. (1996) Stick or Carrot? The Use of Policy Measures to move Waste Management up the Hierarchy. *Waste Management & Research* 14 (4) 385-398.

Winett, R.A., J.H. Kagel, R.C. Battalio and R.C. Winkler (1978) The Effects of Monetary Rebates, Feedback, and Information on Residential Electricity Conservation. *Journal of Applied Psychology* 63: 73-80.

Wittfogel, K. (1957) *Oriental Despotism. A Comparative Study of Total Power.* Yale University Press, New Haven, London.

Wolf, C.P. (1983) Social Impact Assessment: Methodological Overview. in: PADC Enviromental Impact Assessment and Planning Unit: *Environmental Impact Assessement.* pp.253-279. Martinus Nijhoff, Boston/The Hague.

Wolfson, Z. (1994, Boris Komarov) *The Geography of Survival. Ecology in the Post-soviet Era.* M.E.Sharpe, Armonk, New York, London.

Wolsink, M. (1987) Wind Power for the Electricity Supply of Houses. *The Netherlands Journal of Housing and Environmental Research* 2 (3) 195-214.

Wolsink, M. (1989) Attitudes and Expectancies about Wind Turbines and Windfarms. *Wind Engineering* 13 (4) 196-206.

Wolsink, M. (1994) Entanglement of Interests and Motives: Assumptions behind the NIMBY-theory on Facility Siting. *Urban Studies* 31 (6) 851-866.

Wolsink, M. (1996) Dutch Wind Power Policy; Stagnating Implementation of Renewables. *Energy Policy* 24 (12) 1079-1088.

Woo, T.O. and C.H. Castore (1980) Expectancy, Value and Selective Exposure as Determinants of Attitudes toward a Nuclear Power Plant. *Journal of Applied Social Psychology* 10, pp.224-234.

Woodwell, G.M. (1967) Toxic Substances and Ecological Cycles. *Scientific American* 216 (3) 24-31.

Worster, D. (1985) *Rivers of Empire. Water, Aridity, and the Growth of the American West.* Pantheon, New York.

WRR (1984) Netherlands Scientific Council for Government Policy. *Bouwstenen voor een geïntegreerde landbouw.* Report V44, Staatsuitgeverij, The Hague, the Netherlands.

WRR (1992) Netherlands Scientific Council for Government Policy. *Environmental Policy: Strategy, Instruments and Enforcement.* Reports to the Government No.41, Sdu, The Hague, the Netherlands.

Wynne, B. (1987) *Risk Management and Hazardous Wastes: Implementation and the Dialectics of Credibility.* Springer-Verlag, Berlin.

Wynne, B. (1989) Building Public Concern into Risk Management. in: J. Brown (ed.) *Environmental Threats: Perception, Analysis and Management.* Belhaven Press, London/New York. pp.118-132.

Wynne, B. (1989) Frameworks of Rationality in Risk Management: towards the Testing of Naïve Sociology. in: J. Brown (ed.) *Environmental Threats. Perception, Analysis and Management.* Belhaven Press, London/New York. pp.33-47.

Yablokov, A.V. (1990) The Current State of the Soviet Environment. *Environmental Policy Review: The Soviet Union and Eastern Europe.* 4 (1) pp.1-14.

Yanitsky, O. (1993) *Russian Environmentalism: Leading Figures, Facts, Opinions.* Mezhdunarodnyje Otnoshenija Publishing House. Moscow, Russian Republic.

Yearley, S. (1996) *Sociology, Environmentalism, Globalization. Reinventing the Globe*. Sage, London.

Zwick, A (1997) *Global Climate Change: Potential Impact on Human Health.* The ITPS Report no.13, April, pp.27-35.

Subject Index

Author Index

For Product Safety Concerns and Information please contact our EU representative GPSR@taylorandfrancis.com Taylor & Francis Verlag GmbH, Kaufingerstraße 24, 80331 München, Germany

Batch number: 10726526

Printed by Printforce, the Netherlands